Ben Stacy Jerrik (Ed.)

Łazińsk Drugi

Ben Stacy Jerrik (Ed.)

Łazińsk Drugi

Gmina Szamocin, Village, Gmina, Chodzież County, Greater Poland Voivodeship

Part Press

Imprint

Permission is granted to copy, distribute and/or modify this document under the terms of the GNU Free Documentation License, Version 1.2 or any later version published by the Free Software Foundation; with no Invariant Sections, with the Front-Cover Texts, and with the Back- Cover Texts. A copy of the license is included in the section entitled "GNU Free Documentation License".

All parts of this book are extracted from Wikipedia, the free encyclopedia (www.wikipedia.org).

You can get detailed informations about the authors of this collection of articles at the end of this book. The editors (Ed.) of this book are no authors. They have not modified or extended the original texts.

Pictures published in this book can be under different licences than the GNU Free Documentation License. You can get detailed informations about the authors and licences of pictures at the end of this book.

The content of this book was generated collaboratively by volunteers. Please be advised that nothing found here has necessarily been reviewed by people with the expertise required to provide you with complete, accurate or reliable information. Some information in this book maybe misleading or wrong. The Publisher does not guarantee the validity of the information found here. If you need specific advice (f.e. in fields of medical, legal, financial, or risk management questions) please contact a professional who is licensed or knowledgeable in that area.

Any brand names and product names mentioned in this book are subject to trademark, brand or patent protection and are trademarks or registered trademarks of their respective holders. The use of brand names, product names, common names, trade names, product descriptions etc. even without a particular marking in this works is in no way to be construed to mean that such names may be regarded as unrestricted in respect of trademark and brand protection legislation and could thus be used by anyone.

Cover image: www.ingimage.com
Concerning the licence of the cover image please contact ingimage.

Publisher:
Part Press is a trademark of
International Book Market Service Ltd., 17 Rue Meldrum, Beau Bassin, 1713-01 Mauritius
Email: info@bookmarketservice.com
Website: www.bookmarketservice.com

Published in 2011

Printed in: U.S.A., U.K., Germany. This book was not produced in Mauritius.

ISBN: 978-613-8-96090-4

Poland

Republic of Poland
Rzeczpospolita Polska

Anthem:
Mazurek Dąbrowskiego (*Dąbrowski's Mazurka*)

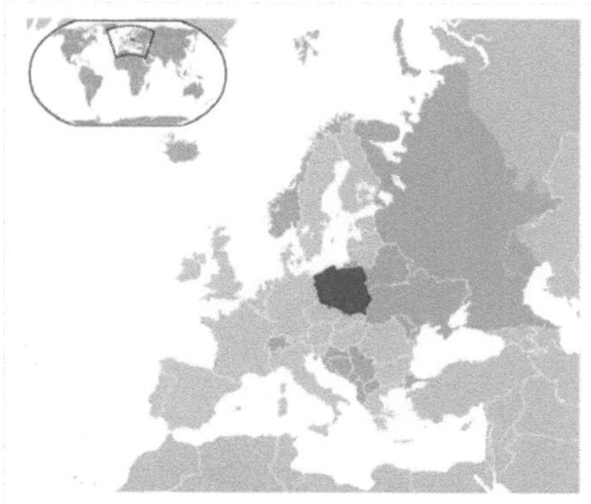

Location of Poland(dark green)

– in Europe(green & dark grey)
– in the European Union(green) — [Legend]

Capital (and largest city)	Warsaw 52°13′N 21°02′E
Ethnic groups (2002)	**96.7% Poles**, 3.3% others and unspecified
Demonym	Pole/Polish
Government	Parliamentary republic
- President	Bronisław Komorowski
- Prime Minister	Donald Tusk
Formation	
- Christianisation[c]	April 14, 966
- First Republic	July 1, 1569
- Second Republic	November 11, 1918
- People's Republic	December 31, 1944
- Third Republic of Poland	January 30, 1990

Area		
-	Total	312685 km² [d](69th) 120,696.41 sq mi
-	Water (%)	3.07
Population		
-	2010 estimate	38,186,860[1] (34th)
-	2002 census	38,231,000
-	Density	120/km² (83rd) 319.9/sq mi
GDP (PPP)		2010 estimate
-	Total	$754,097 billion[2] (20th)
-	Per capita	$19,752[3] (40th)
GDP (nominal)		2010 estimate
-	Total	$468.539 billion[4]
-	Per capita	$12,300[4]
Gini (2002)		34.5
HDI (2011)		▲ 0.813[5] (very high) (39th)
Currency		*Złoty* (PLN)
Time zone		CET (UTC+1)
-	Summer (DST)	CEST (UTC+2)
Drives on the		right
ISO 3166 code		PL
Internet TLD		.pl
Calling code		48
1	**a** See, however, Unofficial mottos of Poland.	
2	**b** Although not official languages, Belarusian, Kashubian, Lithuanian and German are used in 20 communal offices.	
3	**c** The adoption of Christianity in Poland is seen by many Poles, regardless of their religious affiliation or lack thereof, as one of the most significant national historical events; the new religion was used to unify the tribes in the region.	
4	**d** The area of Poland according to the administrative division, as given by the Central Statistical Office, is 312679 km² (sq mi) of which 311888 km² (sq mi) is land area and 791 km² (305 sq mi) is internal water surface area.[6]	

Poland [i]/ˈpoʊlənd/ (Polish: *Polska*), officially the **Republic of Poland** (Polish: *Rzeczpospolita Polska*; Kashubian: *Pòlskô Repùblika*; Silesian: *Polsko Republika*), is a country in Central Europe bordered by Germany to the west; the Czech Republic and Slovakia to the south; Ukraine, Belarus and Lithuania to the east; and the Baltic Sea and Kaliningrad Oblast, a Russian exclave, to the north. The total area of Poland is 312679 square kilometres (sq mi),[6] making it the 69th largest country in the world and the 9th largest in Europe. Poland has a population of over 38 million people,[6] which makes it the 34th most populous country in the world[7] and the sixth most populous member of the European Union, being its most populous post-communist member. Poland is a unitary state made up of sixteen voivodeships. Poland is a member of the European Union, NATO, the United Nations, the World Trade Organization, the Organisation for Economic Co-operation and Development (OECD), European Economic Area,

International Energy Agency, Council of Europe, Organization for Security and Co-operation in Europe, International Atomic Energy Agency and G6.

On 1 July 2011, Poland replaced Hungary as the holder of the rotating Presidency of the Council of the European Union.

The establishment of a Polish state is often identified with the adoption of Christianity by its ruler Mieszko I in 966, over the territory similar to that of present-day Poland. The Kingdom of Poland was formed in 1025, and in 1569 it cemented a long association with the Grand Duchy of Lithuania by signing the Union of Lublin, forming the Polish–Lithuanian Commonwealth. The Commonwealth ceased to exist in 1795 as the Polish lands were partitioned among the Kingdom of Prussia, the Russian Empire, and Austria. Poland regained its independence as the Second Polish Republic in 1918. Two decades later, in September 1939, World War II started with the Nazi Germany and the Soviet Union invasion of Poland. Over six million Polish citizens died in the war. Poland reemerged several years later within the Soviet sphere of influence as the People's Republic in existence until 1989. During the Revolutions of 1989, communist rule was overthrown and soon after, Poland became what is constitutionally known as the "Third Polish Republic".

Despite the vast destruction the country experienced in World War II, Poland managed to preserve much of its cultural wealth. Since the end of the communist period, Poland has achieved a "very high" ranking in terms of human development[8] and standard of living.[9]

Etymology

The source of the name Poland[10] and the ethnonyms for the Poles[11] include endonyms (the way Polish people refer to themselves and their country) and exonyms (the way other peoples refer to the Poles and their country). Endonyms and most exonyms for Poles and Poland derive from the name of the West Slavic tribe of the Polans (*Polanie*). The origin of the name *Polanie* itself is uncertain. It may derive from such Polish words as *pole* (field).[12] The early tribal inhabitants denominated it from the nature of the country. Lowlands and low hills predominate throughout the vast region from the Baltic shores to the foothills of the Carpathian Mountains. *Inter Alpes Huniae et Oceanum est Polonia, sic dicta in eorum idiomate quasi Campania* is the description by Gervase of Tilbury in his *Otia imperialia* (Recreation for the emperor, 1211). In some languages the exonyms for Poland derive from another tribal name, Lechites (*Lechici*).

History

Prehistory

Historians have postulated that throughout Late Antiquity, many distinct ethnic groups populated the regions of what is now known as Poland. The ethnicity and linguistic affiliation of these groups have been hotly debated; the time and route of the original settlement of Slavic peoples in these regions have been the particular subjects of much controversy.

The most famous archeological find from the prehistory and protohistory of Poland is the Biskupin fortified settlement (now reconstructed as a museum), dating from the Lusatian culture of the early Iron Age, around 700 BC. Before adopting Christianity in 960 AD, the people of Poland believed in Svetovid, the Slavic god of war, fertility, and abundance. Many other Slavic nations had the same belief.

Piast dynasty

Baptism of Poland in 966

Poland began to form into a recognizable unitary and territorial entity around the middle of the 10th century under the Piast dynasty. Poland's first historically documented ruler, Mieszko I, was baptized in 966, adopting Catholic Christianity as the nation's new official religion, to which the bulk of the population converted in the course of the next centuries. Played a large role in the Congress of Gniezno. In the 12th century, Poland fragmented into several smaller states when Bolesław divided the nation amongst his sons. In 1226 Konrad I of Masovia, one of the regional Piast dukes, invited the Teutonic Knights to help him fight the Baltic Prussian pagans; a decision which would ultimately lead to centuries of Poland's warfare with the Knights. In 1320, after a number of earlier unsuccessful attempts by regional rulers at uniting the Polish dukedoms, Władysław I consolidated his power, took the throne and became the first King of a reunified Poland. His son, Casimir III, is remembered as one of the greatest Polish kings and is particularly famous for extending royal protection to Jews and providing the original impetus for the establishment of Poland's first university. The Golden Liberty of the nobles began to develop under Casimir's rule, when in return for their military support, the king made serious concessions to the aristocrats, finally establishing their status as superior to that of the townsmen, and aiding their rise to power. When Casimir died in 1370 he left no legitimate male heir and, considering his other male descendants either too young or unsuitable, was laid to rest as the last of the nation's Piast rulers.

Poland was also a centre of migration of peoples. The Jewish community began to settle and flourish in Poland during this era (see History of the Jews in Poland). The Black Death which affected most parts of Europe from 1347 to 1351 did not reach Poland.[13]

Jagiellon dynasty

Nicolaus Copernicus "talking with God", who reveals to him the truth about the solar system. Painting by Jan Matejko.

The rule of the Jagiellon dynasty, spanned the late Middle Ages and early Modern Era of Polish history. Beginning with the Lithuanian Grand Duke Jogaila (Władysław II Jagiełło), the Jagiellon dynasty (1386–1572) formed the Polish–Lithuanian union. The partnership brought vast Lithuania-controlled Rus' areas into Poland's sphere of influence and proved beneficial for the Poles and Lithuanians, who coexisted and cooperated in one of the largest political entities in Europe for the next four centuries. In the Baltic Sea region Poland's struggle with the Teutonic Knights continued and included the Battle of Grunwald (1410), where a Polish-Lithuanian army inflicted a decisive defeat on the Teutonic Knights, both countries' main adversary, allowing Poland's and Lithuania's territorial expansion into the far north region of Livonia.[14] In 1466, after the Thirteen Years' War, King Casimir IV Jagiellon gave royal consent to the milestone Peace of Thorn, which created the future Duchy of Prussia, a Polish vassal. The Jagiellons at one point also established dynastic control over the kingdoms of Bohemia (1471 onwards) and Hungary.[15] [16] In the south Poland confronted the Ottoman Empire and the Crimean Tatars (by whom they were attacked on 75 separate occasions between 1474 and 1569),[17] and in the east helped Lithuania fight the Grand Duchy of Moscow. Some historians estimate that Crimean Tatar slave-raiding cost Poland one million of its population from 1494 to 1694.[18]

Poland was developing as a feudal state, with a predominantly agricultural economy and an increasingly powerful landed nobility. The *Nihil novi* act adopted by the Polish Sejm (parliament) in 1505, transferred most of the legislative power from the monarch to the Sejm, an event which marked the beginning of the period known as "Golden Liberty", when the state was ruled by the "free and equal" Polish nobility. Protestant Reformation movements made deep inroads into Polish Christianity, which resulted in the establishment of policies promoting religious tolerance, unique in Europe at that time. It is believed that this tolerance allowed the country to avoid the religious

Wawel, the seat of Polish kings. Kraków was the nation's capital from 1038 until the move to Warsaw in 1596

turmoil that spread over Western Europe during the late Middle Ages. The European Renaissance evoked in late Jagiellon Poland (kings Sigismund I the Old and Sigismund II Augustus) a sense of urgency in the need to promote a cultural awakening, and resultantly during this period Polish culture and the nation's economy flourished. In 1543 the Pole, Nicolaus Copernicus, an astronomer from Toruń, published his epochal works, *De revolutionibus orbium coelestium* (*On the Revolutions of the Celestial Spheres*), and thus became the first proponent of a predictive mathematical model confirming heliocentric theory which ultimately became the accepted basic model for the practice of modern astronomy. Another major figure associated with the era is classicist poet Jan Kochanowski.[19]

Polish–Lithuanian Commonwealth

The 1569 Union of Lublin established the Polish–Lithuanian Commonwealth, a more closely unified federal state with an elective monarchy, but which was governed largely by the nobility, through a system of local assemblies with a central parliament. The establishment of the Commonwealth coincided with a period of great stability and prosperity in Poland, with the union soon thereafter becoming a great European power and a major cultural entity, occupying approximately one million square kilometres of central Europe, as well as an agent for the expansion of Western culture to the east. Unfortunately it suffered from a number of dynastic crises during the reigns of the Vasa kings Sigismund III and Władysław IV and found itself engaged in a major conflicts with Russia, Sweden and the Ottoman Empire, as well as a series of minor Cossack uprisings.[20]

The Polish–Lithuanian Commonwealth at its greatest extent, after the Truce of Deulino (Dywilino) of 1619

A stylised depiction of King John III Sobieski, allied commander-in-chief and victor at the Battle of Vienna

From the middle of the 17th century, the nobles' democracy, suffering from internal disorder, gradually declined, thus leaving the once powerful Commonwealth extremely vulnerable to foreign intervention. From 1648, the Cossack Khmelnytsky Uprising engulfed the south and east eventually leaving the Ukraine divided, with the eastern part, lost by the Commonwealth, becoming a dependency of the Russian Tsar. This was soon followed by a Swedish invasion, which raged through the Polish heartlands and caused unprecedented damage to infrastructure. Famines and epidemics followed hostilities, and the population dropped from roughly 11 to 7 million.[21] However, under John III Sobieski the Commonwealth's military prowess was re-established, and in 1683 Polish forces played a major part in relieving Vienna of a major Turkish siege which was being conducted by Kara Mustafa in hope of eventually marching his troops further into Europe to spread Islam. Unfortunately, Sobieski's reign was to mark the end of the nation's golden-era, and soon, finding itself subjected to almost constant warfare and suffering enormous population losses as well as massive damage to its economy, the Commonwealth fell into decline. The government became ineffective as a result of large scale internal conflicts (e.g. Lubomirski's Rokosz against John II Casimir and rebellious confederations) and corrupted legislative processes. The nobility fell under the control of a handful of magnates, and this, compounded with two relatively weak kings of the Saxon Wettin dynasty, Augustus II and Augustus III, as well as the rise of Russia and Prussia after the Great Northern War only served to worsen the Commonwealth's plight. Despite this The Commonwealth-Saxony personal union gave rise to the emergence of the Commonwealth's first reform movement, and laid the foundations for the Polish Enlightenment.[22]

During the later part of the 18th century, the Commonwealth made attempts to implement fundamental internal reforms; with the second half of the century bringing a much improved economy, significant population growth and far-reaching progress in the areas of education, intellectual life, art, and especially toward the end of the period, evolution of the social and political system. The most populous capital city of Warsaw replaced Gdańsk (Danzig) as the leading centre of commerce, and the role of the more prosperous townsfolk soon increased. The royal election of 1764 resulted in the elevation of Stanisław August Poniatowski, a refined and worldly aristocrat connected to a major magnate faction, to the monarchy. However, a one-time lover of Empress Catherine II of Russia, the new King spent much of his reign torn between his desire to implement reforms necessary to save his nation, and his perceived necessity to remain in a relationship with his Russian sponsor. This ultimately led to the formation of the 1768 Bar Confederation; a *szlachta* rebellion directed against Russia and the Polish king which fought to preserve Poland's independence and the *szlachta*'s traditional privileges. Unfortunately, attempts at reform provoked the union's neighbours, and in 1772 the First Partition of the

Stanisław August Poniatowski, the last King of Poland and Grand Duke of Lithuania acceded to the throne in 1764, reigning until his abdication on 25 November 1795

Commonwealth by Russia, Austria and Prussia took place; an act which the "Partition Sejm", under considerable duress, eventually "ratified" *fait accompli*.[23] Disregarding this loss, in 1773 the king established the Commission of National Education, the first government education authority in Europe.

The long-lasting Great Sejm convened by Stanisław August in 1788 successfully adopted the May 3 Constitution, the first set of modern supreme national laws in Europe. However, this document, accused by detractors of harbouring revolutionary sympathies, soon generated strong opposition from the Commonwealth's nobles and conservatives as well as from Catherine II, who, determined to prevent the rebirth of a strong Commonwealth set about planning the final dismemberment of the Polish-Lithuanian state. Russia was greatly aided in achieving its goal when the Targowica Confederation, an organisation of Polish nobles, appealed to the Empress for help, and in May 1792 Russian forces crossed the Commonwealth's frontier, thus beginning the Polish-Russian War. The defensive war fought by the Poles and Lithuanians ended prematurely when the King, convinced of the futility of resistance, capitulated and joined the Targowica Confederation. The Confederation then took over the government; but Russia and Prussia, fearing the mere existence of a Polish state, arranged for, and subsequently in 1793, executed the Second Partition of the Commonwealth, which left the country deprived of so much territory that it was practically incapable of independent existence. Eventually, in 1795, following the failed Kościuszko Uprising, the Commonwealth was partitioned one last time by all three of its more powerful neighbours, and with this, effectively ceased to exist.[24]

The Age of Partitions

Tadeusz Kościuszko takes the oath to the King on the Rynek in Kraków, 1794

Poles rebelled several times against the partitioners, particularly near the end of the 18th century and the beginning of the 19th century. One of the most famous and successful attempts at securing renewed Polish independence took place in 1794, during the Kościuszko Uprising, at the Racławice where Tadeusz Kosciuszko, a popular and distinguished general who had served under Washington in America, led peasants and some Polish regulars into battle against numerically superior Russian forces. In 1807, Napoleon I of France recreated a Polish state, the Duchy of Warsaw, but after the Napoleonic Wars, Poland was again divided by the victorious Allies at the Congress of Vienna of 1815. The eastern portion was ruled by the Russian tsar as a Congress Kingdom which possessed a very liberal constitution. However, the tsars soon reduced Polish freedoms, and Russia annexed the country in all but name. Thus in the latter half of the 19th century, only Austrian-ruled Galicia, and particularly the Free City of Kraków, was able to become a centre for Polish cultural life.

Throughout the period of the partitions, political and cultural repression of the Polish nation led to the organisation of a number of uprisings against the authorities of the occupying Russian, Prussian and Austrian governments. Notable amongst these are the November Uprising of 1830 and January Uprising of 1863, both of which were attempts to free Poland from the rule of tsarist Russia. The November uprising began on 29 November 1830 in Warsaw when, led by Lieutenant Piotr Wysocki, young non-commissioned officers at the Imperial Russian Army's military academy in that city revolted. They were soon joined by large segments of Polish society, and together forced Warsaw's Russian garrison to withdraw north of the city.

Over the course of the next seven months, Polish forces successfully defeated the Russian armies of Field Marshal Hans Karl von Diebitsch and a number of other Russian commanders; however, finding themselves in a position unsupported by any other foreign powers, save distant France and the new-born United States, and with Prussia and Austria refusing to allow the import of military supplies through their territories, the Poles accepted that the uprising was doomed to failure. Upon the surrender of Warsaw to General Ivan Paskievich, many Polish troops, feeling they could not go on, withdrew into Germany and there laid down their arms. Poles would have to wait another 32 years for another opportunity to free their homeland.

Polish insurgents and Russian cuirassiers clash on a bridge in Warsaw's Łazienki Park during the November Uprising, Painting by Wojciech Kossak, 1898

When in January 1863 a new Polish uprising against Russian rule began, it did so as a spontaneous protest by young Poles against conscription into the Imperial Russian Army. However, the insurrectionists, despite being joined by high-ranking Polish-Lithuanian officers and numerous politicians were still severely outnumbered and lacking in foreign support. They were forced to resort to guerrilla warfare tactics and ultimately failed to win any major military victories. Afterwards no major uprising was witnessed in the Russian controlled Congress Poland and Poles resorted instead to fostering economic and cultural self-improvement.

Despite the political unrest experienced during the partitions, Poland did benefit from large scale industrialisation and modernisation programs, instituted by the occupying powers, which helped it develop into a more economically coherent and viable entity. This was particularly true in the western provinces annexed by Prussia (later becoming part of the German Empire); an area which eventually, thanks largely to the Greater Poland Uprising was reconstituted as part of the Second Polish Republic and became one of its most productive regions.

Reconstitution of Poland

Chief of State Marshal Józef Piłsudski

During World War I, all the Allies agreed on the reconstitution of Poland that United States President Woodrow Wilson proclaimed in Point 13 of his Fourteen Points. A total of 2 million Polish troops fought with the armies of the three occupying powers, and 450,000 died.[25] Shortly after the armistice with Germany in November 1918, Poland regained its independence as the Second Polish Republic (*II Rzeczpospolita Polska*). It reaffirmed its independence after a series of military conflicts, the most notable being the Polish–Soviet War (1919–1921) when Poland inflicted a crushing defeat on the Red Army at the Battle of Warsaw, an event which is considered to have ultimately halted the advance of Communism into Europe and forced Lenin to rethink his objective of achieving global socialism. Nowadays the event is often referred to as the 'Miracle at the Vistula'.[26]

During this period, Poland successfully managed to fuse the territories of the three former partitioning powers into a cohesive nation state. Railways were restructured to direct traffic towards Warsaw instead of the former imperial capitals, a new network of national roads was gradually built up and a major seaport was opened on the Baltic Coast, so as to allow Polish exports and imports to bypass the politically charged Free City of Danzig.

Poland between 1922 and 1938

The inter-war period heralded in a new era of Polish politics. Whilst Polish political activists had faced heavy censorship in the decades up until the First World War, the country now found itself trying to establish a new political tradition. For this reason, many exiled Polish activists, such as Jan Paderewski (who would later become Prime Minister) returned home to help; a great number of them then went on to take key positions in the newly formed political and governmental structures. Tragedy struck in 1922 when Gabriel Narutowicz, inaugural holder of the Presidency, was assassinated at the Zachęta Gallery in Warsaw by painter and right-wing nationalist Eligiusz Niewiadomski.[27]

The 1926 May Coup of Józef Piłsudski turned rule of the Second Polish Republic over to the Sanacja movement. By the 1930s Poland had become increasingly authoritarian; a number of 'undesirable' political parties, such as the Polish Communists, had been banned and following Piłsudski's death, the regime, unable to appoint a new leader, began to show its inherent internal weaknesses and unwillingness to cooperate in any way with other political parties.

World War II

The Sanacja movement controlled Poland until the start of World War II in 1939, when Nazi Germany invaded on 1 September and the Soviet invasion of Poland followed by breaking the Soviet–Polish Non-Aggression Pact on 17 September. Warsaw capitulated on 28 September 1939. As agreed in the Molotov–Ribbentrop Pact, Poland was split into two zones, one occupied by Germany while the eastern provinces fell under the control of the Soviet Union. By 1941 the Soviets had moved hundreds of thousands of Poles into labor camps all over the Soviet Union, and the Soviet secret police, NKVD, had executed thousands of Polish prisoners of war.[28]

Poland made the fourth-largest troop contribution to the Allied war effort, after the Soviets, the British and the Americans.[a] Polish troops fought under the command of both the Polish Government in Exile in the western theatre of war and under Soviet leadership in the East European theatre. The Polish expeditionary corps, which was controlled by the exiled pre-war government based in London, played an important role in the Italian and North African Campaigns.[29] [30] They are particularly well remembered for their conduct at the Battle of Monte Cassino, a conflict which culminated in the raising of a Polish flag over the ruins of the mountain-top abbey by the 12th Podolian Uhlans. The Polish forces in the West were commanded by Lieutenant General Władysław Anders who had received his command from Prime Minister of the exiled government Władysław Sikorski. On the eastern front, the Soviet-backed Polish 1st Army distinguished itself in the battles for Berlin and Warsaw, although its actions in support of the latter have often been criticised.

Polish forces stationed abroad constituted the fourth largest allied force of the war

At the end of World War II, the gray territories were transferred from Poland to the Soviet Union, and the pink territories from Germany to Poland

Polish servicemen were also active in the theatres of naval and air warfare; during the Battle of Britain Polish squadrons such as the No. 303 "Kościuszko" fighter squadron[31] achieved great success, and by the end of the war the exiled Polish Air Forces could claim 769 confirmed kills. Meanwhile, the navy was active in the protection of convoys in the North Sea and Atlantic Ocean.[32]

In addition to the organised units of the 1st Army and the Forces in the West, the domestic underground resistance movement, the Armia Krajowa *(Home Army)*, fought to free Poland from German occupation and establish an independent Polish state. The wartime resistance movement in Poland was one of the three largest resistance movements of the entire war[b] and encompassed a uniquely large range of clandestine activities, essentially functioning as an underground state complete with underground universities and courts.[33] The resistance was however, largely loyal to the exiled government and generally resented the idea of a communist Poland; for this reason, on 1st August 1944 they initiated Operation Tempest and thus began the Warsaw Uprising.[34] [35] The objective of the uprising was to drive the German occupiers from the city and help with the larger fight against Germany and the Axis powers, however secondary motives for the uprising sought to see Warsaw liberated before the Soviets could reach the capital, so as to underscore Polish sovereignty by empowering the Polish Underground State before the Soviet-backed Polish Committee of National Liberation could assume control. However, a lack of available allied military aid and Stalin's reluctance to allow the 1st Army to help their fellow countrymen take the city, ultimately led to the uprising's failure and subsequent planned destruction of the city.

During the war, German forces, under direct order from Adolf Hitler, set up six major extermination camps, all of which were established on Polish territory; these included both the notorious Treblinka and Auschwitz camps. This allowed the Nazis to transport German Jews outside of 'German' territory, as well as import Jews and other 'undesirables' from across occupied Europe to be 'liquidated' in the concentration camps set up in the General Government. Other undesirables included Polish intelligentsia, Communists, Roma peoples and Soviet Prisoners of War. However, since millions of Jews lived in pre-war Poland, Jewish victims make up the largest percentage of all victims of the Nazis' extermination program. It is estimated that around 90% (or about 3 million members) of pre-war Poland's Jewry were put to death by the Nazis during the Second World War. Throughout the occupation, many members of the Armia Krajowa, supported by the Polish government in exile, and millions of ordinary Poles — at great risk to themselves and their families — engaged in rescuing Jews from the Nazis. Grouped by nationality, Poles represent the largest number of people who rescued Jews during the Holocaust.[36] [37] To date, 6,135 Poles

Grave of Polish fighter killed during the Warsaw Uprising. In the battle, which lasted 63 days, more than 200,000 people died.

have been awarded the title of *Righteous among the Nations* by the State of Israel – more than any other nation.[36] Some estimates put the number of Poles involved in rescue at up to 3 million, and credit Poles with saving up to around 450,000 Jews from certain death.[37]

At the war's conclusion, Poland's borders were shifted westwards, pushing the eastern border to the Curzon Line. Meanwhile, the western border was moved to the Oder-Neisse line. The new Poland emerged 20% smaller by 77500 square kilometres (29900 sq mi). The shift forced the migration of millions of people, most of whom were Poles, Germans, Ukrainians, and Jews.[38] Of all the countries involved in the war, Poland lost the highest percentage of its citizens: over 6 million perished — nearly one-fifth of Poland's population — half of them Polish Jews. Over 90% of the death toll came through non-military losses. Only in the 1970s did Poland again approach its prewar population level.[39] An estimated 600,000 Soviet soldiers died in conquering Poland from German rule.[40]

Postwar communist Poland

Gomułka with Leonid Brezhnev in East Germany

At High Noon, June 4, 1989 – political poster featuring Gary Cooper to encourage votes for the Solidarity party in the 1989 elections. "The 4th of June, 1989 marked a decisive victory for democracy in Poland and, ultimately, across Eastern Europe."—Angela Merkel(English) "Merkel honours Polish freedom struggle and Tiananmen victims". www.thelocal.de. AFP. . Retrieved 2009-09-28.

At the insistence of Joseph Stalin, the Yalta Conference sanctioned the formation of a new Polish provisional and pro-Communist coalition government in Moscow, which ignored the Polish government-in-exile based in London; a move which angered many Poles who considered it a betrayal by the Western Allies. In 1944, Stalin had made guarantees to Churchill and Roosevelt that he would maintain Poland's sovereignty and allow democratic elections to take place; however, upon achieving victory in 1945, the occupying Soviet authorities organised an election which constituted nothing more than a sham and was ultimately used to claim the 'legitimacy' of Soviet hegemony over Polish affairs. The Soviet Union instituted a new communist government in Poland, analogous to much of the rest of the Eastern Bloc. As elsewhere in Eastern Europe the Soviet occupation of Poland met with armed resistance from the outset which continued into the fifties.

Despite widespread objections, the new Polish government accepted the Soviet annexation of the pre-war Eastern regions of Poland[42] (in particular the cities of Wilno and Lwów) and agreed to the permanent garrisoning of Red Army units on Poland's territory. Military alignment within the Warsaw Pact throughout the Cold War came about as a direct result of this change in Poland's political culture and in Western eyes came to characterise the fully-fledged integration of Poland into the brotherhood of communist nations.

The People's Republic of Poland (*Polska Rzeczpospolita Ludowa*) was officially proclaimed in 1952. In 1956, the régime of Władysław Gomułka became temporarily more liberal, freeing many people from prison and expanding some personal freedoms. A similar situation repeated itself in the 1970s under Edward Gierek, but most of the time persecution of anti-communist opposition groups persisted. Despite this, Poland was at the time considered to be one of the least repressive Eastern Bloc states.[43]

Labour turmoil in 1980 led to the formation of the independent trade union "Solidarity" ("*Solidarność*"), which over time became a political force. Despite persecution and imposition of martial law in 1981, it eroded the dominance of the Communist Party and by 1989 had triumphed in Poland's first free and democratic parliamentary elections since the end of the Second World War. Lech Wałęsa, a Solidarity candidate, eventually won the presidency in 1990. The Solidarity movement heralded the collapse of communism across Eastern Europe.

Present-day Poland

Poland joined NATO in 1999 and since 2004 has been a member of the European Union.

A shock therapy programme, initiated by Leszek Balcerowicz in the early 1990s enabled the country to transform its socialist-style planned economy into a market economy. As with all other post-communist countries, Poland suffered temporary slumps in social and economic standards, but it became the first post-communist country to reach its pre-1989 GDP levels, which it achieved by 1995 largely thanks to its booming economy.[44] [45]

Most visibly, there were numerous improvements in human rights, such as the freedom of speech. In 1991, Poland became a member of the Visegrád Group and joined the North Atlantic Treaty Organization (NATO) alliance in 1999 along with the Czech Republic and Hungary. Poles then voted to join the European Union in a referendum in June 2003, with Poland becoming a full member on 1 May 2004. Subsequently Poland joined the Schengen Area in 2007, as a result of which, the country's borders with other member states of the European Union have been dismantled, allowing for full freedom of movement within most of the EU.[46] In contrast to this, the section of Poland's eastern border now comprising the external EU border with Belarus, Russia and Ukraine, has become increasingly well protected, and has led in part to the coining of the phrase 'Fortress Europe', in reference to the seeming 'impossibility' of gaining entry to the EU for citizens of the former Soviet Union.

On April 10, 2010, the President of the Republic of Poland, Lech Kaczyński, along with 89 other high-ranking Polish officials died in a plane crash near Smolensk, Russia. The president's party were on their way to attend an annual service of commemoration for the victims of the Katyń massacre when the tragedy took place.

Geography

Poland's territory extends across several geographical regions, between latitudes 49° and 55° N, and longitudes 14° and 25° E. In the northwest is the Baltic seacoast, which extends from the Bay of Pomerania to the Gulf of Gdańsk. This coast is marked by several spits, coastal lakes (former bays that have been cut off from the sea), and dunes. The largely straight coastline is indented by the Szczecin Lagoon, the Bay of Puck, and the Vistula Lagoon. The centre and parts of the north lie within the North European Plain.

Rising gently above these lowlands is a geographical region comprising the four hilly districts of moraines and moraine-dammed lakes formed during and after the Pleistocene ice age. These lake districts are the Pomeranian Lake District, the Greater Polish Lake District, the Kashubian Lake District, and the Masurian Lake District.

Poland's topography

The Masurian Lake District is the largest of the four and covers much of northeastern Poland. The lake districts form part of the Baltic Ridge, a series of moraine belts along the southern shore of the Baltic Sea.

South of the Northern European Lowlands lie the regions of Silesia and Masovia, which are marked by broad ice-age river valleys. Farther south lies the Polish mountain region, including the Sudetes, the Cracow-Częstochowa Upland, the Świętokrzyskie Mountains, and the Carpathian Mountains, including the Beskids. The highest part of the Carpathians is the Tatra Mountains, along Poland's southern border.

Geology

Granite outcrop *Silesian Rocks/Maiden Rocks* at *Krkonoše Mountains* in southwestern Poland

The geological structure of Poland has been shaped by the continental collision of Europe and Africa over the past 60 million years, on the one hand, and the Quaternary glaciations of northern Europe, on the other. Both processes shaped the Sudetes and the Carpathian Mountains. The moraine landscape of northern Poland contains soils made up mostly of sand or loam, while the ice age river valleys of the south often contain loess. The Cracow-Częstochowa Upland, the Pieniny, and the Western Tatras consist of limestone, while the High Tatras, the Beskids, and the Karkonosze are made up mainly of granite and basalts. The Polish Jura Chain is one of the oldest mountain ranges on earth.

Poland has 70 mountains over 2,000 metres (6,600 ft) in elevation, all in the Tatras. The Polish Tatras, which consist of the High Tatras and the Western Tatras, is the highest mountain group of Poland and of the entire Carpathian range. In the High Tatras lies Poland's highest point, the northwestern peak of Rysy, 2499 metres (8199 ft) in elevation. At its foot lies the mountain lakes Czarny Staw pod Rysami and Morskie Oko.

Giewont in the High Tatras; the mountainous south is a popular destination for hikers

The second highest mountain group in Poland is the Beskids, whose highest peak is Babia Góra, at 1725 metres (5659 ft). The next highest mountain group is the Krkonoše in the Sudetes, whose highest point is Sněžka, at 1602 metres (5256 ft). Among the most beautiful mountains of Poland are the Bieszczady Mountains in the far southeast of Poland, whose highest point in Poland is Tarnica, with an elevation of 1346 metres (4416 ft).

Tourists also frequent the Gorce Mountains in Gorce National Park, with elevations around 1310 metres (4298 ft), and the Pieniny in Pieniny National Park, with elevations around 1050 metres (3445 ft). The lowest point in Poland—at 2 metres (6.6 ft) below sea level—is at Raczki Elbląskie, near Elbląg in the Vistula Delta.

Dunes in Słowiński National Park

Błędów Desert is a desert located in southern Poland in the Silesian Voivodeship and stretches over the Zagłębie Dąbrowskie region. It has a total area of 32 square kilometres (12 sq mi). It is the only desert located in Poland. It is one of only five natural deserts in Europe. It is the warmest desert that appears at this latitude.

It was created thousands of years ago by a melting glacier. The specific geological structure has been of big importance. The average thickness of the sand layer is about 40 metres (131 ft), with a maximum of 70 metres (230 ft), which made the fast and deep drainage very easy.

The sea's activity in Słowiński National Park created sand dunes which in the course of time separated the bay from the Baltic Sea. As waves and wind carry sand inland the dunes slowly move, at a speed of 3 to 10 metres (9.8 to 32.8 ft) meters per year. Some dunes are quite high – up to 30 metres (98 ft). The highest peak of the park — Rowokol (115 metres / 377 feet above sea level) — is also an excellent observation point.

Waters

Vistula River in Modlin

The longest rivers are the Vistula (Polish: *Wisła*), 1047 kilometres (651 mi) long; the Oder (Polish: *Odra*) which forms part of Poland's western border, 854 kilometres (531 mi) long; its tributary, the Warta, 808 kilometres (502 mi) long; and the Bug, a tributary of the Vistula, 772 kilometres (480 mi) long. The Vistula and the Oder flow into the Baltic Sea, as do numerous smaller rivers in Pomerania.

The Łyna and the Angrapa flow by way of the Pregolya to the Baltic, and the Czarna Hańcza flows into the Baltic through the Neman. While the great majority of Poland's rivers drain into the Baltic Sea, Poland's Beskids are the source of some of the upper tributaries of the Orava, which flows via the Váh and the Danube to the Black Sea. The eastern Beskids are also the source of some streams that drain through the Dniester to the Black Sea.

Poland's rivers have been used since early times for navigation. The Vikings, for example, traveled up the Vistula and the Oder in their longships. In the Middle Ages and in early modern times, when the Polish–Lithuanian Commonwealth was the breadbasket of Europe; the shipment of grain and other agricultural products down the Vistula toward Gdańsk and onward to Western Europe took on great importance.

With almost ten thousand closed bodies of water covering more than 1 hectare (2.47 acres) each, Poland has one of the highest number of lakes in the world. In Europe, only Finland has a greater density of lakes. The largest lakes, covering more than 100 square kilometres (39 sq mi), are Lake Śniardwy and Lake Mamry in Masuria, and Lake Łebsko and Lake Drawsko in Pomerania.

Kurtkowiec, oligotrophic lake in southeastern Poland

In addition to the lake districts in the north (in Masuria, Pomerania, Kashubia, Lubuskie, and Greater Poland), there is also a large number of mountain lakes in the Tatras, of which the Morskie Oko is the largest in area. The lake with the greatest depth—of more than 100 metres (328 ft)—is Lake Hańcza in the Wigry Lake District, east of Masuria in Podlaskie Voivodeship.

Holidaymakers relax at the Lake Solina near Lesko in southeastern Poland

Among the first lakes whose shores were settled are those in the Greater Polish Lake District. The stilt house settlement of Biskupin, occupied by more than one thousand residents, was founded before the 7th century BC by people of the Lusatian culture.

Lakes have always played an important role in Polish history and continue to be of great importance to today's modern Polish society. The ancestors of today's Poles, the Polanie, built their first fortresses on islands in these lakes. The legendary Prince Popiel is supposed to have ruled from Kruszwica on Lake Gopło. The first historically documented ruler of Poland, Duke Mieszko I, had his palace on an island in the Warta River in Poznań. Nowadays the Polish lakes provide an invaluable location for the pursuit of water sports such as yachting and wind-surfing.

The Polish Baltic coast is approximately 528 kilometres (328 mi) long and extends from Świnoujście on the islands of Usedom and Wolin in the west to Krynica Morska on the Vistula Spit in the east. For the most part, Poland has a smooth coastline, which has been shaped by the continual movement of sand by currents and winds from west to east. This continual erosion and deposition has formed cliffs, dunes, and spits, many of which have migrated landwards to close off former lagoons, such as Łebsko Lake in Słowiński National Park.

Bay of Puck (*Zatoka Pucka*) in Poland

Prior to the end of the Second World War and subsequent change in national borders, Poland had only a very small coastline; this was situated at the end of the 'Polish Corridor', the only internationally recognised Polish territory which afforded the country access to the sea. However after World War II, the redrawing of Poland's borders and resulting 'shift' of the country to the West left it with a greatly expanded coastline, thus allowing for far greater access to the sea than was ever previously possible. The significance of this event, and importance of it to Poland's future as a major industrialised nation, was allured to by the 1945 Wedding to the Sea.

The largest spits are Hel Peninsula and the Vistula Spit. The largest Polish Baltic island is Wolin. The largest port cities are Gdynia, Gdańsk, Szczecin, and Świnoujście. The main coastal resorts are Sopot, Międzyzdroje, Kołobrzeg, Łeba, Władysławowo, and the Hel Peninsula.

Land use

The patchwork landscape of Masuria

Forests cover 28.8% of Poland's land area. More than half of the land is devoted to agriculture. While the total area under cultivation is declining, the remaining farmland is more intensively cultivated.

More than 1% of Poland's territory, 3145 square kilometres (1214 sq mi), is protected within 23 Polish national parks. Three more national parks are projected for Masuria, the Cracow-Częstochowa Upland, and the eastern Beskids. In addition, wetlands along lakes and rivers in central Poland are legally protected, as are coastal areas in the north. There are over 120 areas designated as landscape parks, along with numerous nature reserves and other protected areas.

Present day Poland is a country with great agricultural prospects; there are over two million private farms in the country, and Poland is the leading producer in Europe of potatoes and rye and is one of the world's largest producers of sugar beets and triticale. This has led Poland to be described on occasion as the future 'bread basket of the European Union'. However, despite employing around 16% of the workforce, agricultural output in Poland remains low and the industry is characterised as largely inefficient due to the large number of small, independent farms. This situation is likely to soon change for the better with the government debating agricultural reform and currently pursuing the option of auctioning off large tracts of state-owned agricultural land.

Biodiversity

Phytogeographically, Poland belongs to the Central European province of the Circumboreal Region within the Boreal Kingdom. According to the World Wide Fund for Nature, the territory of Poland can be subdivided into three ecoregions: the Baltic mixed forests, Central European mixed forests and Carpathian montane conifer forests.

Family of White Stork. Poland hosts the largest White Stork population. [47]

A wisent in the Białowieża Forest

Many animals that have since died out in other parts of Europe still survive in Poland, such as the wisent in the ancient woodland of the Białowieża Forest and in Podlaskie. Other such species include the brown bear in Białowieża, in the Tatras, and in the Beskids, the gray wolf and the Eurasian Lynx in various forests, the moose in northern Poland, and the beaver in Masuria, Pomerania, and Podlaskie.

In the forests, one also encounters game animals, such as Red Deer, Roe Deer and Wild Boars. In eastern Poland there are a number of ancient woodlands, like Białowieża, that have never been cleared by people. There are also large forested areas in the mountains, Masuria, Pomerania, Lubusz Land and Lower Silesia.

Poland is the most important breeding ground for European migratory birds. Out of all of the migratory birds who come to Europe for the summer, one quarter breed in Poland, particularly in the lake districts and the wetlands along the Biebrza, the Narew, and the Warta, which are part of nature reserves or national parks.

Climate

The climate is mostly temperate throughout the country. The climate is oceanic in the north and west and becomes gradually warmer and continental towards the south and east. Summers are generally warm, with average temperatures between 17 °C (63 °F) and 20 °C (68.0 °F). Winters are cold, with average temperatures around 3 °C (37.4 °F) in the northwest and −6 °C (21 °F) in the northeast. Precipitation falls throughout the year, although, especially in the east; winter is drier than summer.

The average daytime summer temperature at sea level along the Baltic coast is 22 °C (71.6 °F) [48]

The warmest region in Poland is Lower Silesian located in south-western Poland where temperatures in the summer average between 22 °C (71.6 °F) and 30 °C (86 °F) but can go as high as 32 °C (89.6 °F) to 38 °C (100.4 °F) on some days in the warmest month of July and August. The warmest cities in Poland are Tarnów, which is situated in Lesser Poland and Wrocław, which is located in Lower Silesian. The average temperatures in Wrocław being 20 °C (68 °F) in the summer and 0 °C (32.0 °F) in the winter, but Tarnów has the longest summer in whole Poland, which lasts for 115 days, from mid-May to mid-September. The coldest region of

Poland is in the northeast in the Podlaskie Voivodeship near the border of Belarus. Usually the coldest city is Suwałki. The climate is affected by cold fronts which come from Scandinavia and Siberia. The average temperature in the winter in Podlaskie ranges from −6 °C (21 °F) to −4 °C (25 °F).

Politics

President of Poland Bronisław Komorowski

Poland is a democracy, with a president as a head of state, whose current constitution dates from 1997. The government structure centers on the Council of Ministers, led by a prime minister. The president appoints the cabinet according to the proposals of the prime minister, typically from the majority coalition in the Sejm. The president is elected by popular vote every five years. The current president is Bronisław Komorowski. Komorowski replaced President Lech Kaczyński following an April 10, 2010 air crash which claimed the life of President Kaczynski, his wife, and 94 other people, during a visit to western Russia for events marking the 70th anniversary of the Katyn massacre. The current prime minister, Donald Tusk, was appointed in 2007 after his party made significant gains in that year's parliamentary elections. In 2011, Tusk became the first Polish prime minister in history to be democratically re-elected for a consecutive term.

Polish voters elect a bicameral parliament consisting of a 460-member lower house (Sejm) and a 100-member Senate (Senat). The Sejm is elected under proportional representation according to the d'Hondt method, a method similar to that used in many parliamentary political systems. The Senat, on the other hand, is elected under the First-past-the-post voting method, with one senator being returned from each of the 100 constituencies.

With the exception of ethnic minority parties, only candidates of political parties receiving at least 5% of the total national vote can enter the Sejm. When sitting in joint session, members of the Sejm and Senat form the National Assembly (the *Zgromadzenie Narodowe*). The National Assembly is formed on three occasions: when a new President takes the oath of office; when an indictment against the President of the Republic is brought to the State Tribunal (*Trybunał Stanu*); and when a president's permanent incapacity to exercise his duties because of the state of his health is declared. To date only the first instance has occurred.

The Sejm's session chamber in Warsaw

The judicial branch plays an important role in decision-making. Its major institutions include the Supreme Court of the Republic of Poland (*Sąd Najwyższy*); the Supreme Administrative Court of the Republic of Poland (*Naczelny Sąd Administracyjny*); the Constitutional Tribunal of the Republic of Poland (*Trybunał Konstytucyjny*); and the State Tribunal of the Republic of Poland (*Trybunał Stanu*). On the approval of the Senat, the Sejm also appoints the ombudsman or the Commissioner for Civil Rights Protection (*Rzecznik Praw Obywatelskich*) for a five-year term. The ombudsman has the duty of guarding the observance and implementation of the rights and liberties of Polish citizens and residents, of the law and of principles of community life and social justice.

In 2011, Poles elected Anna Grodzka as the first ever transsexual MP in European history,[49] and the second transgender MP in European history, after the Italian Vladimir Luxuria.[50]

Law

The Supreme Court building in Warsaw

The Constitution of Poland is the supreme law in contemporary Poland, and the Polish legal system is based on the principle of civil rights, governed by the code of Civil Law. Historically, the most famous Polish legal act is the Constitution of May 3, 1791. Historian Norman Davies describes it as the first of its kind in Europe.[51] The Constitution was instituted as a Government Act (Polish: *Ustawa rządowa*) and then adopted on May 3, 1791 by the Sejm of the Polish–Lithuanian Commonwealth. Primarily, it was designed to redress long-standing political defects of the federative Polish–Lithuanian Commonwealth and its Golden Liberty. Previously only the Henrican articles signed by each of Poland's elected kings could perform the function of a set of basic laws. The new Constitution introduced political equality between townspeople and the nobility (*szlachta*), and placed the peasants under the protection of the government, thus mitigating the worst abuses of serfdom. The Constitution abolished pernicious parliamentary institutions such as the *liberum veto*, which at one time had placed the sejm at the mercy of any deputy who might choose, or be bribed by an interest or foreign power, to have rescinded all the legislation that had been passed by that sejm. The May 3rd Constitution sought to supplant the existing anarchy fostered by some of the country's reactionary magnates, with a more egalitarian and democratic constitutional monarchy.

Unfortunately, the adoption of such a liberal constitution was treated as a grave threat by Poland's more autocratic neighbours. In response Prussia, Austria and Russia formed an anti-Polish alliance and over the next decade collaborated with one another to partition their weaker neighbour and ultimately destroy the Polish state. In the words of two of its co-authors, Ignacy Potocki and Hugo Kołłątaj, the constitution represented "the last will and testament of the expiring Fatherland." Despite this, its text influenced many later democratic movements across the globe.

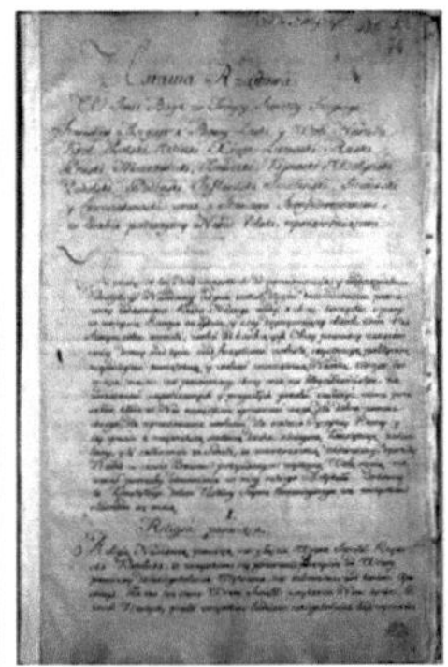

The Constitution of May 3, 1791 has been called the first of its kind in Europe.[51]

Poland's current constitution was adopted by the National Assembly of Poland on 2 April 1997, approved by a national referendum on 25 May 1997, and came into effect on 17 October 1997. It guarantees a multi-party state, the freedoms of religion, speech and assembly, and specifically casts off many Communist ideals to create a 'free market economic system'. It requires public officials to pursue ecologically sound public policy and acknowledges the inviolability of the home, the right to form trade unions, and to strike, whilst at the same time prohibiting the practices of forced medical experimentation, torture and corporal punishment.

Foreign Relations

Foreign Secretary Sikorski meets US Secretary of State Hillary Clinton

In recent years, Poland has extended its responsibilities and position in European and international affairs, supporting and establishing friendly relations with many 'Western' nations and a large number of 'developing' countries.

In 1994, Poland became an associate member of the European Union (EU) and its defensive arm, the Western European Union (WEU), having subimtted preliminary documentation for full membership in 1996, it formally joined the European Union in May 2004, along with the other members of the Visegrád group. In 1996, Poland achieved full OECD membership, and at the 1997 Madrid Summit was invited to join the North Atlantic Treaty Organisation in the first wave of policy enlargement finally becoming a full member of NATO in March 1999.

As changes since the fall of Communism in 1989 have redrawn the map of central Europe, Poland has tried to forge strong and mutually beneficial relationships with its seven new neighbours, this has notably included signing 'friendship treaties' to replace links severed by the collapse of the Warsaw Pact. The Poles have forged special relationships with Lithuania and particularly Ukraine,[52] with whom they will co-host the UEFA Euro 2012 football tournament, in an effort to firmly anchor these states to the 'West' and provide them with an alternative to aligning themselves with the Russian Federation. Despite many positive developments in the region, Poland has found itself in a position where it must seek to defend the rights of ethnic Poles living in the former Soviet Union; this is particularly true of Belarus, where in 2005 the Lukashenko regime launched a campaign against the Polish ethnic minority.[53]

Prime Minister Donald Tusk (right) arrives with former President Lech Wałęsa for the EPP party congress in Warsaw

Poland is the sixth most populous member state of the European Union and, ever since joining in 2004, has pursued policies to increase its role in European affairs. Poland has a grand total of 51 representatives in the European Parliament and in addition to this, since 14 July 2009, former Prime Minister of Poland Jerzy Buzek, has been President of the European Parliament.[54]

Administrative divisions

Poland's current voivodeships (provinces) are largely based on the country's historic regions, whereas those of the past two decades (to 1998) had been centred on and named for individual cities. The new units range in area from less than 10000 square kilometres (3900 sq mi) for Opole Voivodeship to more than 35000 square kilometres (14000 sq mi) for Masovian Voivodeship. Administrative authority at voivodeship level is shared between a government-appointed voivode (governor), an elected regional assembly (*sejmik*) and an executive elected by that assembly.

The voivodeships are subdivided into *powiats* (often referred to in English as counties), and these are further divided into *gminas* (also known as communes or municipalities). Major cities normally have the status of both *gmina* and *powiat*. Poland currently has 16 voivodeships, 379 powiats (including 65 cities with *powiat* status), and 2,478 *gminas*.

Voivodeship		Capital city or cities
in English	*in Polish*	
Greater Poland	*Wielkopolskie*	Poznań
Kuyavian-Pomeranian	*Kujawsko-Pomorskie*	Bydgoszcz / Toruń
Lesser Poland	*Małopolskie*	Kraków
Łódź	*Łódzkie*	Łódź
Lower Silesian	*Dolnośląskie*	Wrocław
Lublin	*Lubelskie*	Lublin
Lubusz	*Lubuskie*	Gorzów Wielkopolski / Zielona Góra
Masovian	*Mazowieckie*	Warsaw
Opole	*Opolskie*	Opole
Podlaskie	*Podlaskie*	Białystok
Pomeranian	*Pomorskie*	Gdańsk
Silesian	*Śląskie*	Katowice
Subcarpathian	*Podkarpackie*	Rzeszów
Świętokrzyskie (Holy Cross)	*Świętokrzyskie*	Kielce
Warmian-Masurian	*Warmińsko-Mazurskie*	Olsztyn
West Pomeranian	*Zachodniopomorskie*	Szczecin

Military

The Polish armed forces are composed of four branches: Land Forces (*Wojska Lądowe*), Navy (*Marynarka Wojenna*), Air Force (*Siły Powietrzne*) and Special Forces (*Wojska Specjalne*). The military is subordinate to the Minister for National Defence, however its sole commander in chief is the President of the Republic.

The Polish army currently consists of 65,000 active personnel, whilst the navy and air force respectively employ 14,300 and 26,126 servicemen and women. The Polish Navy is one of the bigger navies

Polish Air Force F-16 Fighting Falcon

on the Baltic Sea and is mostly involved in Baltic Sea operations such as search and rescue provision for the section of the Baltic under Polish command, as well as hydrographic measurements and research; recently however, the Polish Navy played a more international role as part of the 2003 invasion of Iraq, providing logistical support for the United States Navy. The current position of the Polish Air Force is much the same; it has routinely taken part in Baltic Air Policing assignments, but otherwise, with the exception of a number of units serving in Afghanistan, has seen no active combat since the end of the Second World War. In 2003, the F-16C Block 52 was chosen as the new general multi-role fighter for the air force, the first deliveries taking place in November 2006; it is expected (2010) that the Polish Air Force will create three squadrons of F-16s, which will all be fully operational by 2012.

Polish Army vehicles and troops on patrol in Afghanistan

The most important mission of the armed forces is the defence of Polish territorial integrity and Polish interests abroad.[55] Poland's national security goal is to further integrate with NATO and European defence, economic, and political institutions through the modernisation and reorganisation of its military.[55] Currently the armed forces is being re-organised according to NATO standards, and as of 1 January 2010, the transition to an entirely contract-based military has been completed. Previously male citizens were expected to complete a period of active service with the military; since 2007 up until the amendment of the law on conscription, the obligatory term of service was nine months.[56]

Polish military doctrine reflects the same defensive nature as that of its NATO partners. From 1953 to 2009 Poland was a large contributor to various United Nations peacekeeping missions.[55] [57] The Polish Armed Forces took part in the 2003 invasion of Iraq, deploying 2,500 soldiers in the south of that country and commanding the 17-nation Multinational force in Iraq.

The military was temporarily, but severely, affected by the loss of many of its top commanders in the wake the 2010 Polish Air Force Tu-154 crash near Smolensk, Russia, which killed all 96 passengers and crew, including, amongst others, the Chief of the Polish Army's General Staff Franciszek Gągor and Polish Air Force commanding general Andrzej Błasik. They were en route from Warsaw to attend an event to mark the 70th anniversary of the Katyn massacre, whose site is commemorated approximately 19 km west of Smolensk.[58] [59]

Law enforcement and emergency services

Poland has a highly developed system of law enforcement with a long history of effective policing by the State Police Service. The structure of law enforcement agencies within Poland is a multi-tier one, with the State Police providing criminal-investigative services, Municipal Police serving to maintain public order and a number of other specialised agencies, such as the Polish Border Guard, acting to fulfil their assigned missions. In addition to these state services, private security companies are also common, although they possess no powers assigned to state agencies, such as, for example, the power to make an arrest or detain a suspect.

A mounted officer of the State Police in Poznań

Emergency services in Poland consist of the Emergency Medical Services, Search and Rescue units of the Polish Armed Forces and State Fire Service. Emergency medical services in Poland are, unlike other services, provided for by local and regional government.

Since joining the European Union all of Poland's emergency services have been undergoing major restructuring and have, in the process, acquired large amounts of new equipment and staff.[60] All emergency services personnel are now uniformed and can be easily recognised thanks to a number of innovative design features, such as reflective paint and printing, present throughout their service dress and vehicle liveries. In addition to this, in an effort to comply with EU standards and safety regulations, the police and other agencies have been steadily replacing and modernising their fleets of vehicles; this has left them with thousands of new automobiles, as well as many new aircraft, boats and helicopters.[61]

Economy

Poland's high-income economy[62] is considered to be one of the healthiest of the post-Communist countries and is currently one of the fastest growing within the EU. Having a strong domestic market, low private debt, flexible currency, and not being dependent on a single export sector, Poland is the only European economy to have avoided the late-2000s recession.[63] Since the fall of the communist government, Poland has steadfastly pursued a policy of liberalising the economy and today stands out as a successful example of the transition from a centrally planned economy to a primarily market-based economy. In 2009 Poland had the highest GDP growth in the EU. As of November 2009, the Polish economy has not entered the global recession of the late 1st decade of the 21st century nor has it even contracted.[64] [65]

Financial centre of Warsaw

The privatization of small and medium state-owned companies and a liberal law on establishing new firms have allowed the development of an aggressive private sector. As a consequence, consumer rights organizations have also appeared. Restructuring and privatisation of "sensitive sectors" such as coal, steel, rail transport and energy has been continuing since 1990. Between 2007 and 2010, the government plans to float twenty public companies on the Warsaw Stock Exchange, including parts of the coal industry. The biggest privatisations have been the sale of the national telecoms firm Telekomunikacja Polska to France Télécom in 2000, and an issue of 30% of the shares in Poland's largest bank, PKO Bank Polski, on the Polish stockmarket in 2004.

Poland is part of the Schengen Area and the EU single market, and it is among the countries obliged to join the Eurozone (green) (the Eurozone itself is blue).

The Polish banking sector is the largest in central and eastern Europe as well being as the largest and the most highly developed sector of the country's financial markets. It is regulated by the Polish Financial Supervision Authority. During the transformation to a market-oriented economy, the government privatized some banks, recapitalized the rest and introduced legal reforms that made the sector competitive. This has attracted a significant number of strategic foreign investors. Poland's banking sector has approximately 5 domestic banks, a network of nearly 600 cooperative banks and 18 branches of foreign-owned banks. In addition, foreign investors have controlling stakes in nearly 40 commercial banks, which make up 68% of the banking capital.[66]

Poland has a large number of private farms in its agricultural sector, with the potential to become a leading producer of food in the European Union. Structural reforms in health care, education, the pension system, and state administration have resulted in larger-than-expected fiscal pressures. Warsaw leads Central Europe in foreign investment.[67] GDP growth had been strong and steady from 1993 to 2000 with only a short slowdown from 2001 to 2002.

The economy had growth of 3.7% annually in 2003, a rise from 1.4% annually in 2002. In 2004, GDP growth equaled 5.4%, in 2005 3.3% and in 2006 6.2%.[68] According to Eurostat data, Polish PPS GDP per capita stood at 61% of the EU average in 2009.[69]

Although the Polish economy is currently undergoing economic development, there are many challenges ahead. The most notable task on the horizon is the preparation of the economy (through continuing deep structural reforms) to allow Poland to meet the strict economic criteria for entry into the Eurozone. According to the Polish foreign minister Radosław Sikorski the country could join the eurozone before 2016.[70] Some businesses may already accept the euro as payment. In addition, the ability to establish and conduct business easily has been cause for economic hardship as the World Economic Forum recently ranked Poland near the bottom of OECD countries in terms of the clarity, efficiency and neutrality of its legal framework for firm to settle disputes.[71] A report concluded that on-going foreign business

Gdynia, situated at Gdańsk Bay on the south coast of the Baltic Sea, is an important seaport of Poland.

disputes issues may "have damaged Poland's reputation as an attractive location for FDI" by reinforcing the impression of "Poland's substandard reputation for maintaining an efficient and neutral framework to settle business disputes involving multinational foreign investors."[72] Ernst & Young's 2010 European attractiveness survey reported that Poland saw a 52% decrease in FDI job creation and a 42% decrease in number of FDI projects since 2008.[73]

Average salaries in the enterprise sector in December 2010 were 3,848 PLN (1,012 euro or 1,374 US dollars)[74] and growing sharply.[75] Salaries vary between the regions: the median wage in the capital city Warsaw was 4,603 PLN (1,177 euro or 1,680 US dollars) while in Kielce it was only 3,083 PLN (788 euro or 1125 US dollars). Differences in salaries in various districts of Poland is even higher and range from 2,020 PLN (517 euro or 737 US dollars) in Kępno County, which is located in Greater Poland Voivodeship to 5,616 (1,436 euro or 2,050 US dollars) in Lubin County, which lies in Lower Silesian Voivodeship.[76]

According to a Credit Suisse report, Poles are the second wealthiest (after Czechs) of the Central European peoples.[77] [78] This makes Poland an attractive destination for many guest workers from Asia and Eastern Europe. Even though Poland is rather an ethnically homogeneous country, the number of foreigners is growing every year.[79]

Since the United Kingdom, Ireland and some other European countries opened their job markets for Poles, many workers, especially from rural regions, have left the country to seek a better wages abroad. However, there is a rapid growth of the salaries, booming economy, strong value of Polish currency, and quickly decreasing unemployment (from 14.2% in May 2006 to 6.7% in August 2008).[80] Commodities produced in Poland include: electronics, cars (including the luxurious Leopard car), buses (Autosan, Solaris, Solbus), helicopters (PZL Świdnik), transport equipment, locomotives, planes (PZL Mielec), ships, military engineering (including tanks, SPAAG systems), medicines (Polpharma, Polfa), food, clothes, glass, pottery (Bolesławiec), chemical products and others.

Corporations

Poland is recognised as a regional economic power within Central Europe, possessing nearly 40 percent of the 500 biggest companies in the region (by revenues).[81] Poland was the only member of the EU to avoid the recession of the late 2000s, a testament to the Polish economy's stability.[82] The country's most competitive firms are components of the WIG20 which is traded on the Warsaw Stock Exchange.

Well known Polish brands include, amongst others, Tyskie, LOT Polish Airlines, PKN Orlen, E. Wedel, Empik, Poczta Polska, PKO Bank, PKP, Mostostal, PZU Insurance, and TVP.[83]

Poland is recognised as having an economy with significant development potential, overtaking the Netherlands in mid-2010 to become Europe's sixth largest economy.[84] Foreign Direct Investment in Poland has remained strong ever since the country's re-democratisation following the Round Table Agreement in 1989. Despite this, problems do exist, and further progress in achieving success depends largely on the government's privatisation of Poland's remaining state industries and continuing development and modernisation of the economy.

The list includes the largest companies by turnover in 2009, but does not include major banks or insurance companies:

The Warsaw Stock Exchange is, by market capitalisation, one of Central Europe's largest

Warsaw is home to many of Poland's largest business enterprises

Rank in 2009[85]	Name of concern	Location of headquarters	Revenue (Thou. PLN)	Profit (Thou. PLN)	Employees
1.	PKN Orlen SA	Płock	49,876,074	1,907,812	4,482
2.	PGE SA	Lublin	22,418,183	5,378,534	46,357
3.	Fiat Auto Poland SA	Bielsko-Biała	19,788,511	820,420	6,421
4.	PGNiG SA	Warsaw	19,493,756	1,442,103	31,685
5.	PBG SA	Poznań	17,761,936	N/A	27,032
6.	Metro Group Poland	Warsaw	16,800,000	N/A	24,077
7.	Telekomunikacja Polska SA	Warsaw	16,560,000	1,597,000	27,667
8.	Lotos Group SA	Gdańsk	15,379,754	1,091,950	4,949
9.	Tauron Group SA	Katowice	13,859,135	1,165,020	28,839
10.	KGHM Polska Miedź SA	Lubin	12,286,107	3,070,799	18,370

Tourism

Poland is a major part of the global tourism market and is currently experiencing an upward trend in its number of visitors; this began shortly after joining the European Union.[86] Tourism in Poland contributes to the country's overall economy and makes up a relatively large proportion of the country's service market. The most attractive urban destinations for tourists are Warsaw, Kraków, Wrocław, Poznań, Lublin and Toruń; in addition to these the historic site of the Auschwitz concentration camp near Oświęcim is a noteworthy place of pilgrimage and a now constitutes a major monument to the prevention of war and suffering in Southern Poland. Popular areas of natural beauty include northeast Poland's Masurian Lake District and

The city of Zamość is a UNESCO World Heritage Site and one of the main tourist attractions of Lublin Voivodeship.

Białowieża Forest. Poland's main tourist offerings are thought to be based around city-sightseeing and extra-urban historical monuments, business trips, qualified tourism, agrotourism, and mountain hiking, among others.

Poland was the 17th most visited country by foreign tourists in 2008.[87]

Energy

Żerań power station in Warsaw

The electricity generation sector in Poland is still largely fossil-fuel based. Many power plants nationwide use Poland's position as a major European exporter of coal to their advantage by continuing to use coal as the primary raw material in production of their energy; in 2007, hard bituminous coal contributed 48% of energy generation, brown coal and gas 12% each and oil 23%.[88] Currently the three largest Polish coal mining firms (*Weglokoks, Kompania Węglowa and JSW*) extract around 100 million tonnes of coal annually; all three of these companies are key constituents of the Warsaw Stock Exchange's lead economic indexes.

Renewable forms of energy currently only account for a small proportion of Poland's full energy generation capacity.[89] However, the national government has set targets for the development of renewable energy sources in Poland which should see the portion of power produced by renewable resources climb to 7.5% by 2010 and 15% by 2020. This is to be achieved mainly through the construction of wind farms and a number of hydroelectric stations.

Poland is thought to have around 164,800,000,000 m³ of proven natural gas reserves and around 96,380,000 barrels of proven oil reserves. These reserves are currently attended to and exploited by energy supply companies such as PKN Orlen (*the only Polish company listed in the Fortune Global 500*). However, due to the small amounts of fossil fuels naturally occurring in Poland not being enough to satisfy the full energy consumption needs of the population and thus need to buy from abroad, the country is considered to be a net importer of oil and natural gas.

Transport

Today transport in Poland is provided by means of rail, road, shipping and air travel. Positioned in Central Europe and with an eastern border compromising the largest external border of the Schengen Area with the rest of Eastern Europe, Poland has long been, and remains a key country through which imports to and exports from the European Union pass.

LOT Boeing 767 'City of Poznań' on stand at Warsaw Chopin Airport

Since joining the EU in 2004, Poland has invested large amounts of money into the modernisation of its transport networks. The country now has a developing expressways network compromised of motorways such as the A4 and express roads such as the S7. In addition to these newly built roads, many local and regional roads are being rebuilt as part of a national programme to rebuild all roads in Poland.[90]

The A4 Motorway near Kraków

Again, with regard to railways, much the same situation is taking place. The Polish authorities have begun a program by which they hope to increase operating speeds across the entire Polish rail network; this is particularly true of a number of national trunk routes which are expected to soon receive new rolling stock capable of speeds over 200 km/h. Finally, there is a plan to introduce high speed rail to Poland from around 2014. The Polish government recently revealed that it intends to connect all major cities to a future high-speed rail network by 2020.[91] Most intercity rail operations in Poland are operated by PKP Intercity whilst regional trains are run by a number of operators, the largest of which is Przewozy Regionalne.

The air and maritime transport markets in Poland are largely well developed, although construction of new airport and seaport facilities is ongoing. Poland has a number of international airports; the largest of which is Warsaw Chopin Airport, the primary global hub for LOT Polish Airlines, which is the largest airline of Eastern Central Europe and one of the world's oldest airlines still in operation today. Seaports exist all along Poland's Baltic Sea coast, with most freight operations using either Gdynia or Gdańsk as their base. Passenger ferries link Poland with Scandinavia all year round; these services are provided from Gdańsk by Polferries, Stena Line from Gdynia and Unity Line from the Port of Świnoujście.

Science

In 1925, the chemist Maria Skłodowska-Curie established the first Radium Institute in Poland.[92]

According to Frost & Sullivan's Country Industry Forecast the country is becoming an interesting location for research and development investments.[93] Multinational companies such as: ABB, Delphi, GlaxoSmithKline, Google, Hewlett–Packard, IBM, Intel, LG Electronics, Microsoft, Motorola, Siemens and Samsung have set up research and development centres in Poland.[94] Over 40 research and development centers and 4,500 researchers make Poland the biggest research and development hub in Central and Eastern Europe.[93] Companies chose Poland because of the availability of highly qualified labor force, presence of universities, support of authorities, and the largest market in Central Europe.[93]

Today Poland's tertiary education institutions; traditional universities (found in its major cities), as well as technical, medical, and economic institutions, employ around 61,000 researchers and members of staff. There are around 300 research and development institutes, with about 10,000 researchers. In total, there are around 91,000 scientists in Poland today. However, in the 19th and 20th centuries many Polish scientists worked abroad; one of the greatest of these exiles was Maria Skłodowska-Curie, a physicist and chemist who lived much of her life in France. In the first half of the 20th century, Poland was a flourishing centre of mathematics. Outstanding Polish mathematicians formed the Lwów School of Mathematics (with Stefan Banach, Hugo Steinhaus, Stanisław Ulam) and Warsaw School of Mathematics (with Alfred

Warsaw's Staszic Palace is home to the Polish Academy of Sciences

Tarski, Kazimierz Kuratowski, Wacław Sierpiński). The events of World War II pushed many of them into exile. Such was the case of Benoît Mandelbrot, whose family left Poland when he was still a child. An alumnus of the Warsaw School of Mathematics was Antoni Zygmund, one of the shapers of 20th-century mathematical analysis.

According to a KPMG report[95] 80% of Poland's current investors are content with their choice and willing to reinvest. In 2006, Intel decided to double the number of employees in its research and development centre in Gdańsk.[94]

Communications

The Main Municipal Post Office of Bydgoszcz

The share of the telecom sector in the GDP is 4.4% (end of 2000 figure), compared to 2.5% in 1996. Nevertheless, despite high expenditures for telecom infrastructure (the coverage increased from 78 users per 1,000 inhabitants in 1989 to 282 in 2000).

The value of the telecommunication market is zl 38.2bn (2006), and it grew by 12.4% in 2007 PMR.[96] The coverage mobile cellular is over 1000 users per 1000 people (2007). Telephones—mobile cellular: 38.7 million (Onet.pl & GUS Report, 2007), telephones—main lines in use: 12.5 million (Telecom Team Report, 2005).

With regard to internet access, the most popular ADSL services for home users in Poland are Neostrada provided by TPSA, and Net24 provided by Netia. Business users as well as some home users use Internet DSL TP also offered by TPSA. According to Eurostat, OECD and others, Internet access in Poland is amidst the most expensive in Europe. This is mostly caused by the lack of competitiveness. New operators, such as Dialog and GTS Energis are making their own provider lines and offer more attractive and cheaper service. Recently, the Polish Office of Electronical Communication passed a bill forcing the TPSA to rent 51% of their ADSL lines to other ISPs for 60% lower prices. This move will definitely affect the prices of DSL in Poland.

TP S.A. headquarters in Warsaw

The public postal service in Poland is operated by Poczta Polska (The Polish Post). It was created on October 18, 1558, when king Zygmunt August established a permanent postal route from Kraków to Venice (later also to Wilno) in order to manage affairs in Italy that arose after the death of Queen Bona, his mother. Since then the service was dissolved on a number of occasions, most notably during the partitions of Poland. After regaining independence in 1918, the united territory of Poland was in need of a uniform network of communication. Thus, the interwar period saw the rapid development of the postal system as new services were introduced (e.g. money transfers, payment of pensions, delivery of magazines, air mail). Although during national uprisings and in the course of wars communication was provided mainly through field post, which was subject to military authority, postmen always took active part in the fight for independence by secretly delivering parcels and documents, or by providing vital information about the enemy. Many important events in the history of Poland involved the postal service, like the heroic Defence of the Polish Post Office in Gdańsk in 1939 and the participation of the Polish Scouts' Postal Service in the Warsaw Uprising. During the difficult times of the Second World War, the Polish Post in exile would lift up the spirits of compatriots by issuing postage stamps. Nowadays the service is a modern, functioning state-owned company which provides a number of standard and express delivery options, as well as operating the Polish postal home-delivery service. The postal service is currently expanding into the provision of logistical services.

Demographics

Poland, with 38,116,000 inhabitants,[6] has the eighth-largest population in Europe and the sixth-largest in the European Union. It has a population density of 122 inhabitants per square kilometer (328 per square mile).

Long Market in Gdańsk filled with picturesque Dutch style tenements is a favourite meeting place in the Kashubian capital.

Poland historically contained many languages, cultures and religions on its soil. The country had a particularly large Jewish population prior to World War II, when the Nazi Holocaust caused Poland's Jewish population, estimated at 3 million before the war, to drop to just 300,000. The outcome of the war, particularly the westward shift of Poland's borders to the area between the Curzon Line and the Oder-Neisse line, coupled with post-war expulsion of minorities, significantly reduced the country's ethnic diversity. Over 7 million Germans fled or were expelled from the Polish side of the Oder-Neisse boundary.[97]

According to the 2002 census, 36,983,700 people, or 96.74% of the population, consider themselves Polish, while 471,500 (1.23%) declared another nationality, and 774,900 (2.03%) did not declare any nationality. The largest minority nationalities and ethnic groups in Poland are Silesians (about 200,000), Germans (152,897 according to the census, 92% in Opole Voivodeship and Silesian Voivodeship), Belarusians (c. 49,000), Ukrainians (c. 30,000), Lithuanians, Russians, Roma, Jews, Lemkos, Slovaks, Czechs, and Lipka Tatars.[98] Among foreign citizens, the Vietnamese are the largest ethnic group, followed by Greeks and Armenians.

The Polish language, part of the West Slavic branch of the Slavic languages, functions as the official language of Poland. Until recent decades Russian was commonly learned as a second language but has been replaced by English and German as the most common second languages studied and spoken.[99]

In recent years, Poland's population has decreased because of an increase in emigration and a sharp drop in the birth rate. Since Poland's accession to the European Union, a significant number of Poles have emigrated to Western European countries such as the United Kingdom, Germany and Ireland in search of work. Some organizations have stated that Polish emigration is primarily caused by Poland's high

The Main Market Square in Kraków is the heart of Poland's southern cultural capital

unemployment rate (10.5% in 2007), with Poles searching for better work opportunities abroad. In April 2007, the Polish population of the United Kingdom had risen to approximately 300,000, and estimates place the Polish population in Ireland at 65,000. Some sources claim that the number of Polish citizens who emigrated to the UK after 2004 is as high as 2 million.[100] This, however, is contrasted by a recent trend that shows that more Poles are entering the country than leaving it.[101]

Polish minorities are still present in the neighboring countries of Ukraine, Belarus, and Lithuania, as well as in other countries (see Poles for population numbers). Altogether, the number of ethnic Poles living abroad is estimated to be around 20 million.[102] The largest number of Poles outside of Poland can be found in the United States.[103]

Urbanization

	Rank	City	Voivodeship	Population	Metropolitan area	
Warsaw **Kraków** **Łódź**	1	**Warsaw**	Masovian	1,716,855 ▲	2,680,600	**Wrocław** **Poznań** **Gdańsk**
	2	**Kraków**	Lesser Poland	755,546 ▲	1,449,700	
	3	**Łódź**	Łódź	739,832 ▼	1,061,600	
	4	**Wrocław**	Lower Silesian	632,561 ▲	1,136,900	
	5	**Poznań**	Greater Poland	552,735 ▼	943,700	
	6	**Gdańsk**	Pomeranian	456,874 ▲	1,080,700	
	7	**Szczecin**	West Pomeranian	405,944 ▼	683,900	
	8	**Bydgoszcz**	Kuyavian-Pomeranian	356,936 ▼	724,700	
	9	**Lublin**	Lublin	348,961 ▼	465,900	
	10	**Katowice**	Silesian	307,699 ▼	3,239,200	
	11	**Białystok**	Podlaskie	294,989 ▲	370,000	
	12	**Gdynia**	Pomeranian	247,539 ▼	1,080,700	
	13	**Częstochowa**	Silesian	238,737 ▼	400,000	
	14	**Radom**	Masovian	223,110 ▼	371,000	
	15	**Sosnowiec**	Silesian	218,422 ▼	3,239,200	

Religion

Until World War II, Poland was a religiously diverse society, in which substantial Jewish, Protestant and Christian Orthodox minorities coexisted with a Roman Catholic majority. As a result of the Holocaust and the post–World War II flight and expulsion of German and Ukrainian populations, Poland has become overwhelmingly Roman Catholic. In 2007, 88.4% of the population belonged to the Catholic Church.[104] Though rates of religious observance are lower, at 52%[105] to 60%,[106] Poland remains one of the most devoutly religious countries in Europe.[107]

Basilica of Our Lady of Licheń

Holy Spirit Orthodox Church in Białystok

Religious minorities include Polish Orthodox (about 506,800),[6] various Protestants (about 150,000),[6] Jehovah's Witnesses (126,827),[6] Eastern Catholics, Mariavites, Polish Catholics, Jews, and Muslims (including the Tatars of Białystok). Members of Protestant churches include about 77,500 in the largest Evangelical-Augsburg Church,[6] and a similar number in smaller Pentecostal and Evangelical churches.

From 16 October 1978 until his death on 2 April 2005 Karol Józef Wojtyła (later Pope John Paul II), a natural born Pole, reigned as Supreme Pontiff of the Roman Catholic Church and Sovereign of

Vatican City. His was the second-longest documented pontificate; only Pope Pius IX served longer. He has been the only Slavic and Polish Pope to date, and was the first non-Italian Pope since Dutch Pope Adrian VI in 1522.[108] Additionally he is credited with having played a significant role in hastening the downfall of communism in Poland and throughout Eastern Europe; he is famously quoted as having, at the height of communism in 1979, told Poles "not be afraid", later praying: "Let your Spirit descend and change the image of the land... this land".[109] [110] He is a deeply revered figure within Polish society, and his passing in 2005 was met with large-scale outpourings of national grief.

Freedom of religion is now guaranteed by the 1989 statute of the Polish Constitution,[111] enabling the emergence of additional denominations.[112] However, because of pressure from the Polish Episcopate, the exposition of doctrine has entered the public education system as well.[113] [114] According to a 2007 survey, 72% of respondents were not opposed to religious instruction in public schools; alternative courses in ethics are available only in one percent of the entire public educational system.[115]

Famous sites of Christian pilgrimage in Poland include the Monastery of Jasna Góra in the southern Polish city of Częstochowa, as well as the Family home of John Paul II in Wadowice just outside of Kraków.

Health

Poland's healthcare system is based on an all-inclusive insurance system. State subsidised healthcare is available to all Polish citizens who are covered by this general health insurance program. However, it is not compulsory to be treated in a state-run hospital as a number of private medical complexes do exist nationwide.[116]

All medical service providers and hospitals in Poland are subordinate to the Polish Ministry of Health, which provides oversight and scrutiny of general medical practice as well as being responsible for the day to day administration of the healthcare system. In addition to these roles, the ministry is also tasked with the maintenance of standards of hygiene and patient-care.

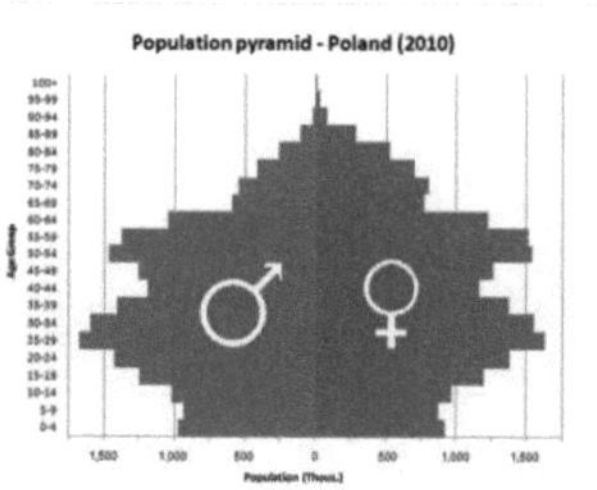

Poland's population pyramid for 2010

Main clinical building of the Gdańsk University Medical Centre in Wrzeszcz

Hospitals in Poland are organised according to the regional administrative structure, resultantly most towns have their own hospital *(Szpital Miejski)*. Larger and more specialised medical complexes tend only to be found in major cities, with some even more specialised units located only in the capital, Warsaw. However, all voivodeships have their own general hospital (most have more than one), all of which are obliged to have a trauma centre; these types of hospital, which are able to deal with almost all medical problems are called 'regional hospitals' *(Szpital Wojewódzki)*. The last category of hospital in Poland is that of specialised medical centres, an example of which would be the Skłodowska-Curie Institute of Oncology, Poland's leading, and most highly specialised centre for the research and treatment of cancer.

The Polish health-care industry is currently undergoing a major transformation, with many hospitals being listed as top priorities for refurbishment.[117] As a result of this process, many hospitals have already been thoroughly modernised throughout and are now equipped with the latest in medical hardware. The overall quality of healthcare provision nationwide, as judged by European standards, is generally regarded as being very high.[118] This is reflected in the nation's average life expectancy, which at 71 for males and 80 for females,[119] has shown a marked increase from 63/68 in 2003, and now corresponds with the average figures for life expectancy in the European

Union.

Education

The education of Polish society was a goal of rulers as early as the 12th century, and Poland soon became one of the most educated countries in Europe. The library catalogue of the Cathedral Chapter of Kraków dating back to 1110 shows that in the early 12th century Polish intellectuals had access to European literature. The Jagiellonian University, founded in 1364 by King Casimir III in Kraków, is one of Europe's oldest universities. In 1773 King Stanisław August Poniatowski established the Commission of National Education (*Komisja Edukacji Narodowej*), the world's first state ministry of education.

The *Collegium Maius* is the oldest building of the Jagiellonian University in Kraków

The first university in Poland, Kraków's Jagiellonian University, was established in 1364 by Casimir III the Great in Kraków. It is the oldest university in Poland. It is the second oldest university in Central Europe and one of the oldest universities in the world. The idea to found the university was first conceived when Poland's King Casimir III realized that the nation needed a class of educated people, especially lawyers, who could codify the country's laws and administer the courts and offices. His efforts to found an institution of higher learning in Poland were finally rewarded when Pope Urban V granted him permission to open the University of Krakow.

The wearing of traditional academic dress is an important feature of Polish educational ceremonies

Since changes made in 2009 education in Poland starts at the age of five or six for the 0 class (Kindergarten) and six or seven years in the 1st class of primary school (Polish *szkoła podstawowa*). It is compulsory that children do one year of formal education before entering 1st class at no later than 7 years of age. At the end of 6th class when the students are 13, they take a compulsory exam that will determine to which lower secondary school (*gimnazjum, pronounced gheem-nah-sium*) (Middle School/Junior High) they will be accepted. They will attend this school for three years for classes, 7, 8, and 9. They then take another compulsory exam to determine the upper secondary level school they will attend. There are several alternatives, the most common being the three years in a *liceum* or four years in a technikum. Both end with a maturity examination (matura, quite similar to French baccalauréat), and may be followed by several forms of upper education, leading to licencjat or inżynier (the Polish Bologna Process first cycle qualification), magister (the Polish Bologna Process second cycle qualification) and eventually doktor (the Polish Bologna Process third cycle qualification).[120]

There are currently 18 fully accredited traditional universities in Poland, these are then further supplemented by 20 technical universities, nine independent medical universities and five universities for the study of economics. In addition to these institutions there are then nine agricultural academies, three pedagogical universities, a theological academy and three maritime service universities. Poland's long and history of promoting the arts has led to the establishment of a number of higher educational institutes dedicated to the teaching of the arts. Amongst these are the seven higher state academies of music. All of these institutions are further supplemented by a large number of private educational institutions and the four national military

The Adam Mickiewicz University, Poznań

academies (two for the army and one for each of the other branches of service), bringing the total number of organisations for the pursuit of higher education to well over 500, one of the largest numbers in Europe. The Programme for International Student Assessment, coordinated by the OECD, currently ranks Poland's educational system as the 23rd best in the world, being neither significantly higher nor lower than the OECD average.[121]

Culture

The culture of Poland is closely connected with its intricate 1000 year history[122] Its unique character developed as a result of its geography at the confluence of Western and Eastern Europe. With origins in the culture of the Proto-Slavs, over time Polish culture has been profoundly influenced by its interweaving ties with the Germanic, Latinate and Byzantine worlds as well as in continual dialog with the many other ethnic groups and minorities living in Poland.[123] The people of Poland have traditionally been seen as hospitable to artists from abroad and eager to follow cultural and artistic trends popular in other countries. In the 19th and 20th centuries the Polish focus on cultural advancement often took precedence over political and economic activity. These factors have contributed to the versatile nature of Polish art, with all its complex nuances.[123]

Famous people

Poland is the birthplace of many distinguished personalities (see. List of Polish people), among which are: Mikołaj Kopernik,[124] Fryderyk Chopin,[125] [126] Maria Skłodowska Curie,[127] Tadeusz Kościuszko, Kazimierz Pułaski, Józef Piłsudski and Pope John Paul II. Great polish painter Jan Matejko devoted his monumental art to the most significant historical events on Polish lands, along with the playwright, painter and poet Stanisław Wyspiański. Stanisław Ignacy Witkiewicz (Witkacy) was an example of a Polish avant-garde philosopher and author of aesthetic theories.

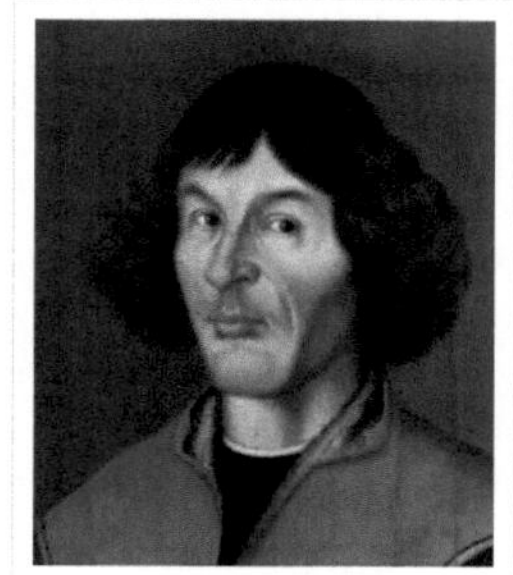
Mikołaj Kopernik
(Latin: *Nicolaus Copernicus*)

Polish literature dates back to the 12th century[128] and includes many famous poets and writers such as Jan Kochanowski, Adam Mickiewicz, Bolesław Prus, Juliusz Słowacki, Witold Gombrowicz, Stanisław Lem and, Ryszard Kapuściński. Writers Henryk Sienkiewicz, Władysław Reymont, Czesław Miłosz, Wisława Szymborska have each won the Nobel Prize in Literature. Also a renowned Polish novelist, who wrote in the English language, was Joseph Conrad.[129]

Many world famous Polish movie directors include Academy Awards winners Roman Polański, Andrzej Wajda, Zbigniew Rybczyński, Janusz Kamiński, Krzysztof Kieślowski, Agnieszka Holland. World renowned actresses were Helena Modjeska and Pola Negri.

Poles have outstanding achievements in mountaineering in the Himalayas. The most famous Polish climbers are Jerzy Kukuczka, Krzysztof Wielicki, Piotr Pustelnik, Andrzej Zawada and Wanda Rutkiewicz.

Society

The Grand Hotel in Sopot. Poland's Baltic Sea resorts are very popular tourist destinations amongst Poles as well as foreign tourists

Poland has a great, long standing tradition of tolerance towards minorities, as well as absence of discrimination on the grounds of religion, nationality or race. It has a high level of gender equality, promotes disability rights movement, and is legally and socially tolerant towards homosexuality. Poland is the first country in the world where corporal punishment was prohibited.[130] Poland has, throughout most of its long history, experienced only very limited immigration from abroad; this trend can largely be attributed to Poland's lack of slavery and overseas colonies as well as its lack of existence as a state during much of the 19th and early 20th centuries. Despite this, the country has for a long time been regarded as having a very tolerant society, which affords equal rights to all people no matter what their ethnic background. This can be said to stem largely from the reign of king Casimir III the Great and his acceptance for Poland's Jewish community, in a time when the most of Europe recessed antisemitic mood. The history of Jews in Poland shows peaceful co-existence of a nation and particular ethnic group.

Pope John Paul II is considered to have been a great promoter of Poland around the world

As many as 96.7% of Polish citizens declare to be Poles, and 97.8% declare that they speak Polish at home (Census 2002). The population of Poland became one of the most ethnically homogeneous in the world as a result of the radically altered borders after World War II and the subsequent migrations. This homogeneity is a result of post World War II deportations ordered by the Soviet Union authorities, who wished to remove the sizeable Polish minorities from Lithuania, Belarus and Ukraine and repatriation of Ukrainians from Poland to the Soviet Union (see territorial changes of Poland and historical demography of Poland for details). Unlike in many other countries, the minority rights in Poland are guaranteed directly in the Constitution of Poland (art. 35), and today there are, amongst others, sizeable German, Ukrainian and Belarusian minorities present in the country.[131]

After the formal collapse of Communism in 1989, Poland greatly improved its image in the world and thus has received further support from the country's recent economic success and effective entry into the structures of the European Union. Polish citizens have obtained a good reputation as workers in the united Europe, mainly due to the broad range of jobs beyond the borders of their state, since 2004. The results of an OSCE survey from 2004 showed that Poles work the second most hours per week of any nationality worldwide.

Music

Fryderyk Chopin, composer

Artists from Poland, including famous composers like Chopin or Penderecki and traditional, regionalized folk musicians, create a lively and diverse music scene, which even recognizes its own music genres, such as poezja śpiewana and disco polo. As of 2006, Poland is one of the few countries in Europe where rock and hip hop dominate over pop music, while all kinds of alternative music genres are encouraged.

The origins of Polish music can be traced as far back as the 13th century; manuscripts have been found in Stary Sącz, containing polyphonic compositions related to the Parisian Notre Dame School. Other early compositions, such as the melody of *Bogurodzica* and *Bóg się rodzi* (a coronation polonaise for Polish kings by an unknown composer), may also date back to this period, however, the first known notable composer, Mikołaj z Radomia, was born and lived in the 15th century. During the 16th century, two main musical groups – both based in Kraków and belonging to the King and Archbishop of the Wawel – led to the rapid development of Polish music. Composers writing during this period include Wacław z Szamotuł, Mikołaj Zieleński, and Mikołaj Gomółka. Diomedes Cato, a native-born Italian who lived in Kraków from about the age of five, became one of the most famous lutenists at the court of Sigismund III, and not only imported some of the musical styles from southern Europe, but blended them with native folk music.[132]

At the end of the 18th century, Polish classical music evolved into national forms like the polonaise. In the 19th century the most popular composers were: Józef Elsner and his pupils Fryderyk Chopin and Ignacy Dobrzyński. Important opera composers of the era were Karol Kurpiński and Stanisław Moniuszko whilst the list of famous soloists and composers included Henryk Wieniawski, Juliusz Zarębski. At the turn of the 19th and 20th centuries the most promiment composers could said to have been Władysław Zeleński and Mieczysław Karłowicz, with Karol Szymanowski gaining prominence prior to World War II. Alexandre Tansman lived in Paris but had strong connections with Poland. Henryk Górecki and Krzysztof Penderecki composed in Poland, Andrzej Panufnik emigrated.

Tomasz Stańko is a popular contemporary Polish jazz musician

Traditional Polish folk music has had a major effect on the works of many well-known Polish composers, and no more so than on Fryderyk Chopin, a widely recognised national hero of the arts. All of Chopin's works involve the piano and are technically demanding, emphasising nuance and expressive depth. As a great composer, Chopin invented the musical form known as the instrumental ballade and made major innovations to the piano sonata, mazurka, waltz, nocturne, polonaise, étude, impromptu and prélude, he was also the composer of a number of polonaises which borrowed heavily from traditional Polish folk music. It is largely thanks to him that the such pieces gained great popularity throughout Europe during the 19th century. Nowadays the most distinctive folk music can be heard in the towns and villages of the mountainous south, particularly in the region surrounding the winter resort town of Zakopane.

Today Poland has a very active music scene, with the jazz and metal genres being particularly popular amongst the contemporary populace. Polish jazz musicians such as Krzysztof Komeda, created a unique style, which was most famous in 60s and 70s and continues to be popular to this day. Since the fall of Communism, Poland has become a major venue for large-scale music festivals, chief among which are the Open'er Festival, Opole Festival and Sopot Festival.

Media

Headquarters of national daily Gazeta Wyborcza in Warsaw

Poland has instituted freedom of press since the fall of communism, a system under which the media was heavily politically controlled and censored. However, public TV and radio are still regulated by the government, this is exercised through an agency called *Krajowa Rada Radiofonii i Telewizji* (*The National Radio and Television Committee*), which is similar to television regulatory commissions in other developed nations.

Poland has a number of major media outlets, chief amongst which are the national television channels. TVP is Poland's public broadcasting corporation; about a third of its income comes from a broadcast receiver licence, while the rest is made through revenue from commercials and sponsorships. State television operates two mainstream channels, TVP 1 and TVP 2, as well as regional programs (TVP Info) for each of the country's 16 voivodeships. In addition to these general channels, TVP runs a number of genre-specific programmes such as TVP Sport, TVP Historia, TVP Kultura and TVP Seriale; there are currently plans to run channels dedicated to the coverage of political affairs (TVP Parlament) and entertainment (TVP Rozrywka).

Poland has a number of internationally broadcast and 24 hour news channels, chief amongst which are Polsat News, TVN 24, and TV Polonia, the latter is a state-run channel dedicated to the transmission of Polish language television for the Polish diaspora abroad. There are a number of major private television outlets such as Polsat and the TVN network.

Studio set of TVP Wrocław's 'Fakty' news program

Poland has a highly developed printed news industry, with daily newspapers like Gazeta Wyborcza *(The Electoral Gazette)* and Rzeczpospolita *(The Republic)* providing more traditional, intellectually stimulating reporting and tabloids such as Fakt providing more sensationalist writing which is less current affairs orientated. Rzeczpospolita is one of the nation's oldest publications still in operation today, founded in 1920, it has become a stalwart bastion of Polish reporting and in 2006 won a prestigious award for being, along with the Guardian (a British daily), the best designed newspaper in the world.[133] In early 2005, Rzeczpospolita found itself at the very centre of a heated public debate, after one of its employees, the former dissident and journalist Bronisław Wildstein, abstracted a list with the names of 240,000 informers and victims of the communist secret police from the Institute of National Remembrance and distributed it among colleagues. In the wake of the incident, Wildstein was dismissed from the paper's staff. This event represents one of the most controversial episodes in the history of the modern Polish media; which is largely due to the ongoing dispute over whether the names of communist era agents and collaborators should be disclosed or not.

Major media outlets are experiencing an ongoing restructuring which is seeing many of them amalgamated into major media groups; a prime example of which is the German Axel Springer AG Publishing conglomerate's purchase of Fakt. International cooperation is also a growing trend within Polish media; TVP recently began cooperating with the French-German TV network ARTE.

Literature

Polish literature has a long and complicated history. During the Middle Ages most Polish authors and academics (Jan Długosz) wrote only in Latin, as at the time, this was the 'academic' language which linked Europe together; Jan Kochanowski broke this trend and became the first author to write the majority of his works in the Polish language. A number of Polish authors have won great renown in the past few centuries, however, this largely stems from the initial success of the works of Adam Mickiewicz, who wrote the first Polish epic, Pan Tadeusz, in 1834. Influential authors of the late 19th and 20th centuries include Henryk Sienkiewicz, Władysław Reymont, Witold Gombrowicz and Czesław Miłosz. To date four Polish authors have won the nobel prize for literature, with Władysław Reymont being one of only nine writers to receive the prestigious award for one particular, outstanding literary work *(awarded for the great national epic, 'The Peasants' in 1924)* rather than their career as a whole.[134] Ferdynand Antoni Ossendowski was an internationally popular writer.

Adam Mickiewicz (1798–1855)	Henryk Sienkiewicz (1846–1916)	Władysław Reymont (1865–1925)	Witold Gombrowicz (1904–1969)	Czesław Miłosz (1911–2004)

With regard to poetry, Poland has a long and distinguished history of producing world-class poets. Chief among these are the 'three bards' (trzej wieszcze), Mickiewicz, Krasiński and Słowacki; the three national poets of Polish Romantic literature. Incidentally, the Polish word *Wieszcz* means 'prophet' or 'soothsayer', a fitting reference for the three, as the bards were thought to not only voice Polish national sentiments but to frequently foresee the nation's future.

Today the traditions of Polish literature and poetry are being carried forward by a new generation of writers. Included within this group is Wisława Szymborska, a best-selling author and recipient of the 1996 Nobel Prize in Literature.[135]

Architecture

Further information: Category:Polish architecture

The Gothic St Mary's Basilica on the Main Market Square in Kraków

Polish cities and towns reflect the whole spectrum of European styles. Romanesque architecture is represented by St. Andrew's Church in Kraków, and characteristic for Poland Brick Gothic by St. Mary's Church in Gdańsk. Richly decorated attics and arcade loggias are the common elements of the Polish Renaissance architecture,[136] [137] like in City Hall in Poznań. For some time the late renaissance, so called mannerism, most notably in Bishop's Palace in Kielce, coexisted with the early baroque like in Church of SS. Peter and Paul in Kraków.

History has not been kind to Poland's architectural monuments. Nonetheless, a number of ancient structures had survived: castles, churches, and stately buildings, often unique in the regional or European context. Some of them have been painstakingly restored, like Wawel Castle, or completely reconstructed after being destroyed in the Second World War, including the Old Town and Royal Castle in Warsaw and the Old Town of Gdańsk. The architecture of Gdańsk is mostly of the Hanseatic variety, a Gothic style common amongst the former trading cities along the Baltic sea and in the northern part of Central Eastern Europe. The architectural style of Wrocław is mainly representative of German architecture, since it was for centuries located within the German states. The centre of Kazimierz Dolny on the Vistula is a good example of a well-preserved medieval town. Poland's ancient capital, Kraków, ranks among the best-preserved Gothic and Renaissance urban complexes in Europe. Meanwhile, the legacy of the Kresy Marchlands of Poland's eastern regions, where Wilno and

Renaissance City Hall in Poznań

Lwów (now *Vilnius* and *Lviv*) were recognised as two major centres for the arts, played a special role in the development of Polish architecture, with Catholic church architecture deserving special note.[123] In Vilnius (Lithuania) there are about 40 baroque and Renaissance churches. In Lviv (Ukraine) there are also a number of Gothic, Renaissance, and baroque religious buildings which have borrowed from and been influenced by Orthodox and Armenian church architecture.

The second half of the 17th century is marked by baroque architecture. Side towers, visible in Branicki Palace in Białystok are typical for Polish baroque. The classical Silesian baroque is represented by the University in Wrocław. Profuse decorations of Branicki Palace in Warsaw are characteristic of rococo style. The centre of Polish classicism was Warsaw under the rule of the last Polish king Stanisław August Poniatowski.[138] The Palace on the Water is the most notable example of Polish neoclassical architecture. Lublin Castle represents the Gothic Revival style in architecture, while the Izrael Poznański Palace in Łódź is an example of eclecticism.

Sports

The Tall Ships' Races 2007, Szczecin

Many sports are popular in Poland. Football (soccer) is the country's most popular sport, with a rich history of international competition.[139] [140] Track and field, basketball, boxing, ski jumping, fencing, handball, ice hockey, swimming, volleyball, and weightlifting are other popular sports. The golden era of football in Poland occurred throughout the 1970s and went on until the early 1980s when the Polish national football team achieved their best results in any FIFA World Cup competitions finishing 3rd place in the 1974 and 1982 editions. The team won a gold medal in football at the 1972 Summer Olympics and also won two silver medals in 1976 and 1992. Poland, along with Ukraine, will host the UEFA European Football Championship in 2012.[141]

The Polish men's national volleyball team is ranked 5th in the world and the women's volleyball team is ranked 10th. Mariusz Pudzianowski is a highly successful strongman competitor and has won more World's Strongest Man titles than any other competitor in the world, winning the event in 2008 for the fifth time. The first Polish Formula One driver, Robert Kubica, has brought awareness of Formula One Racing to Poland. Poland has made a distinctive mark in motorcycle speedway racing thanks to Tomasz Gollob, a highly successful Polish rider. The national speedway team of Poland is one of the major teams in international speedway and is very successful in various competitions.[142]

The Polish mountains are an ideal venue for hiking, skiing and mountain biking and attract millions of tourists every year from all over the world.[87] Baltic beaches and resorts are popular locations for fishing, canoeing, kayaking and a broad-range of other water-themed sports.

Cuisine

Polish cuisine has influenced the cuisines of its surrounding countries. For centuries the Polish kitchen has been the arena for competing with France and Italy. It is rich in meat, especially chicken and pork, and winter vegetables (cabbage in the dish bigos), and spices, as well as different kinds of pasta the most notable of which are the pierogi. Polish national cuisine shares some similarities with other Central European and Eastern European traditions. Generally speaking, Polish cuisine is hearty. The preparation of traditional cuisine generally is time intensive and Poles allow themselves a generous amount of time to prepare and enjoy their festive meals, with some meals (like Christmas Eve or Easter breakfast) taking a number of days to prepare

Kotlet schabowy served with potatoes.

in their entirety. It is worth noting that most regions of Poland have their own local gastronomic traditions and distinctive flavours.[143]

Notable foods in Polish cuisine include: soups - rosół, barszcz, żurek, krupnik, kapuśniak, zupa pomidorowa, zupa ogórkowa, zupa grzybowa, flaczki (tripe soup); pierogi, kiełbasa, gołąbki, oscypek, kotlet schabowy, kotlet mielony, bigos, various potato dishes, kanapka, and many more. Traditional Polish desserts include pączki, faworki, gingerbread, babka and others.

Characteristic dishes are soured milk, buttermilk, kefir, gherkin, pickled cucumber, sauerkraut.

International rankings

The following are links to international rankings of Poland.

Index	Rank	Countries reviewed
Human Development Index 2010	41st	169
Corruption Perceptions Index 2010	41st	178
OECD Working time	2nd	27
Index of Economic Freedom 2011	68th	157
Privacy International Yearly Privacy ranking of countries, 2007	19th	45
Reporters Without Borders Press Freedom Index 2010	32nd	178
UNICEF Child Well-being league table [144]	14th	21
Networked Readiness Index 2010–2011	62nd	138
OICA Automobile Production 2010	18th	48

See also

- Polish British
- Polish American

Notes

a Numerous sources state that Polish Army was the Allies fourth biggest fighting contingent. Steven J. Zaloga and Richard Hook write that "by the war's end the Polish Army was the fourth largest contingent of the Allied coalition after the armed forces of the Soviet Union, the United States and the United Kingdom."[145] Jerzy Jan Lerski writes "All in all, the Polish units, although divided and controlled by different political orientation, constituted the fourth largest Allied force, after the America, British and Soviet Armies."[146] M. K. Dziewanowski has noted that "if Polish forces fighting in the east and west were added to the resistance fighters, Poland had the fourth largest Allied army in the war (after the USSR, the U.S. and Britain)".[147]

The claim of the fourth biggest Ally need to be taken in perspective, throughout the war Poland's position varied from the 2nd biggest Ally (after the fall of France, when Polish army outnumbered the French) to perhaps the 5th at the end (after the USA, Soviet Union, China and Britain). For details, see the analysis in Polish contribution to World War II.

b Sources vary with regards to what was the largest resistance movement during World War II. The confusion often stems from the fact that as war progressed, some resistance movements grew larger – and other diminished. Polish territories were mostly freed from Nazi German control in the years 1944–1945, eliminating the need for their respective (anti-Nazi) partisan forces (in Poland (although the cursed soldiers continued to fight against the Soviets). Several sources note that Polish Armia Krajowa was the largest resistance movement in Nazi-occupied Europe. For example, Norman Davies wrote "Armia Krajowa (Home Army), the AK, which could fairly claim to be the largest of European resistance";[148] Gregor Dallas wrote "Home Army (Armia Krajowa or AK) in late 1943 numbered around 400000, making it the largest resistance organization in Europe";[149] Mark Wyman wrote "Armia Krajowa was considered the largest underground resistance unit in wartime Europe".[150] Certainly, Polish resistance was the largest resistance till German invasion of Yugoslavia and invasion of the Soviet Union in 1941. After that point, the numbers of Soviet partisans and Yugoslav partisans begun growing rapidly. The numbers of Soviet partisans quickly caught up and were very similar to that of the Polish resistance.[151] [152] The numbers of Tito's Yugoslav partisans were roughly similar to those of the Polish and Soviet partisans in the first years of the war (1941–1942), but grew rapidly in the latter years, outnumbering the Polish and Soviet partisans by 2:1 or more (estimates give Yugoslavian forces about 800,000 in 1945, to Polish and Soviet forces of 400,000 in 1944).[152] [153]

References

[1] "GUS – Population as of 30.06.2010" (http://www.stat.gov.pl/gus/5840_655_ENG_HTML.htm). Stat.gov.pl. . Retrieved 2011-05-26.

[2] "Gross domestic product (2010)" (http://siteresources.worldbank.org/DATASTATISTICS/Resources/GDP_PPP.pdf). *The World Bank: World Development Indicators database*. World Bank. 1 July 2011. . Retrieved 2011-07-04.

[3] Data refer to the year 2009 and 2010. GDP (PPP) (http://siteresources.worldbank.org/DATASTATISTICS/Resources/GDP_PPP.pdf) & Population (http://siteresources.worldbank.org/DATASTATISTICS/Resources/POP.pdf), World Development Indicators database, World Bank. Accessed on July 7, 2011.

[4] "Poland" (http://www.imf.org/external/pubs/ft/weo/2011/01/weodata/weorept.aspx?sy=2008&ey=2011&scsm=1&ssd=1& sort=country&ds=.&br=1&c=964&s=NGDPD,NGDPDPC,PPPGDP,PPPPC,LP&grp=0&a=&pr.x=59&pr.y=12). International Monetary Fund. . Retrieved 2011-05-06.

[5] "Table 1 – Human Development Index and its components" (http://hdr.undp.org/en/media/HDR_2010_EN_Table1.pdf). *Human Development Index 2010*. Human Development Reports. . Retrieved 4 November 2010.

[6] "Concise Statistical Yearbook of Poland, 2008" (http://www.stat.gov.pl/cps/rde/xbcr/gus/PUBL_maly_rocznik_statystyczny_2008.pdf) (PDF). Central Statistical Office (Poland). 28 July 2008. . Retrieved 2008-08-12.

[7] NationMaster.com 2003–2007, Poland, Facts and figures (http://www.nationmaster.com/country/pl-poland)

[8] (English) "Human Development Index and its components" (http://hdr.undp.org/en/media/HDR_2010_EN_Table1.pdf). *hdr.undp.org*. . Retrieved 2011-08-27.

[9] "Interactive Infographic of the World's Best Countries" (http://www.newsweek.com/2010/08/15/
interactive-infographic-of-the-worlds-best-countries.html). Newsweek. 15 August 2010. . Retrieved 28 July 2011.

[10] **Polenia** by Thietmar of Merseburg *Chronicle*, 1002. (German: *Polen*)

[11] **Polani** by John Canaparius, *Vita sancti Adalberti episcopi Pragensis*, or *Life of St. Adalbert of Prague*, 999.

[12] "fr. pal, pele, altd. pal, pael, dn. pael, sw. pale, isl. pall, bre. pal, peul, it. polo, pole, pila, [in:] A dictionary of the Anglo-Saxon languages.
Joseph Bosworth. S.275.; planus, plain, flat; from Indo- Germanic pele, flat, to spread, also the root of words like plan, floor, and field. [in:]
John Hejduk. Soundings. 1993. p. 399"; "the root pele is the source of the English words "field" and "floor". The root "plak" is the source of
the English word "flake" [in:] Loren Edward Meierding. Ace the Verbal on the SAT. 2005. p. 82

[13] Teeple, J. B. (2002). *Timelines of World History*. Publisher: DK Adult.

[14] Jerzy Wyrozumski – *Historia Polski do roku 1505* (History of Poland until 1505), Państwowe Wydawnictwo Naukowe (Polish Scientific
Publishers PWN), Warszawa 1986, ISBN 978-83-01-03732-1

[15] (**English**) Norman Davies (1996). *Europe: a history*. Oxford University Press. p. 428. ISBN 01-98201-71-0. "By 1490 the Jagiellons
controlled Poland-Lithuania, Bohemia, and Hungary, but not the Empire."

[16] " Jagiellon dynasty (European history) (http://www.britannica.com/EBchecked/topic/299331/Jagiellon-dynasty)". Encyclopædia
Britannica.

[17] Davies (2007). *Warfare, State and Society on the Black Sea Steppe,1500–1700*. (http://books.google.com/books?id=XH4hghHo1qoC&
pg=PA17&dq&hl=en#v=onepage&q=&f=false). p.17.

[18] " The Crimean Tatars and their Russian-Captive Slaves (http://www.econ.hit-u.ac.jp/~areastd/mediterranean/mw/pdf/18/10.pdf)"
(PDF). Eizo Matsuki, *Mediterranean Studies Group at Hitotsubashi University*.

[19] Józef Andrzej Gierowski – *Historia Polski 1505–1764* (History of Poland 1505–1764), Państwowe Wydawnictwo Naukowe (Polish
Scientific Publishers PWN), Warszawa 1986, ISBN 978-83-01-03732-1

[20] Józef Andrzej Gierowski – *Historia Polski 1505–1764* (History of Poland 1505–1764), p. 105-173

[21] " Poland – The 17th-century crisis (http://www.britannica.com/EBchecked/topic/466681/Poland/28190/The-17th-century-crisis)".
Britannica Online Encyclopedia.

[22] Józef Andrzej Gierowski – *Historia Polski 1505–1764* (History of Poland 1505–1764), p. 174-301

[23] Józef Andrzej Gierowski – *Historia Polski 1764–1864* (History of Poland 1764–1864), Państwowe Wydawnictwo Naukowe (Polish
Scientific Publishers PWN), Warszawa 1986, ISBN 978-83-01-03732-1, p. 1-74

[24] Józef Andrzej Gierowski – *Historia Polski 1764–1864* (History of Poland 1764–1864), p. 74-101

[25] " Poland – Independence Won and Lost, 1914–45. (http://countrystudies.us/poland/13.htm)". Library of Congress Country Studies.

[26] Frątczak, Sławomir Z. (2005). "Cud nad Wisłą" (http://web.archive.org/web/20070708173639/http://www.glos.com.pl/
Archiwum_nowe/Rok+2005/032/strona/Cud.html) (in Polish). *Głos* (32/2005). Archived from the original (http://www.glos.com.pl/
Archiwum_nowe/Rok+2005/032/strona/Cud.html) on 2007-07-08. . Retrieved June 18, 2006.

[27] Bitter glory: Poland and its fate, 1918 to 1939; p.179

[28] " Russian parliament condemns Stalin for Katyn massacre (http://www.bbc.co.uk/news/world-europe-11845315)". BBC News.
November 26, 2010

[29] At the siege of Tobruk

[30] including the capture of the monastery hill at the Battle of Monte Cassino

[31] (**English**) Lynne Olson & Stanley Cloud. 2003. A Question of Honor. The Kosciuszko Squadron: Forgotten Heroes of World War II. New
York: Knopf.

[32] Peszke, Michael Alfred (February 1999). *Poland's Navy, 1918–1945*. Hippocrene Books. p. 37. ISBN 978-0-7818-0672-5.

[33] Stanisław Salmonowicz, *Polskie Państwo Podziemne*, Wydawnictwa Szkolne i Pedagogiczne, Warszawa, 1994, ISBN 978-83-02-05500-3,
p.37

[34] The Warsaw Rising (http://www.polandinexile.com/rising.htm), polandinexile.com

[35] Richard J. Kozicki, Piotr Wróbel (eds), *Historical Dictionary of Poland, 966–1945*, Greenwood Press, 1996, ISBN 978-0-313-26007-0,
Google Print, p.34 (http://books.google.com/books?id=S6aUBuWPqywC&pg=PA34&ots=VwP7o88St6&dq=Zygmunt+Berling+
Warsaw+Uprising&sig=CuHYk98fb5OPXX7ZTyEiFKnuc5I)

[36] Yad Vashem, The Holocaust Martyrs' and Heroes' Remembrance Authority, Righteous Among the Nations – per Country & Ethnic Origin
January 1, 2009. Statistics (http://www1.yadvashem.org/righteous_new/statistics.html)

[37] Richard C. Lukas, *Out of the Inferno: Poles Remember the Holocaust*, University Press of Kentucky 1989 – 201 pages. Page 13; also in
Richard C. Lukas, *The Forgotten Holocaust: The Poles Under German Occupation, 1939–1944*, University Press of Kentucky, 1986, Google
Print, p.13 (http://books.google.ca/books?id=lz9obsxmuW4C&pg=PA13&dq="&sig=ACfU3U0SGgyvqSbL4bypepYoO_CbYc_N_w).

[38] " European Refugee Movements After World War Two (http://www.bbc.co.uk/history/worldwars/wwtwo/refugees_01.shtml)". BBC
– History.

[39] " Poland – population (http://countrystudies.us/poland/27.htm)". Library of Congress Country Studies.

[40] Abbott Gleason (2009). " *A Companion to Russian History* (http://books.google.com/books?id=JyN0hlKcfTcC&pg=PA409&dq&
hl=en#v=onepage&q=&f=false)". Wiley-Blackwell. p.409. ISBN 1-4051-3560-3

[41] (**English**) "Merkel honours Polish freedom struggle and Tiananmen victims" (http://www.thelocal.de/politics/20090604-19717.html).
www.thelocal.de. AFP. . Retrieved 2009-09-28.

[42] Arthur Bliss Lane *I saw Poland betrayed: An American Ambassador Reports to the American People*. Indianapolis: The Bobbs-Merrill Company, 1948.

[43] (Polish) Polska. Historia (http://encyklopedia.pwn.pl/haslo.php?id=4575043) PWN Encyklopedia. Retrieved 11 July 2005.

[44] "Real GDP growth in CEECs" (http://transitioneconomies.blogspot.com/2006/05/real-gdp-growth-in-ceecs.html). Transitioneconomies.blogspot.com. 2006-05-28. . Retrieved 2009-05-06.

[45] "WHY POLAND?" (http://polisci.lsa.umich.edu/documents/jjackson/chapt1.pdf) (PDF). . Retrieved 2009-07-08.

[46] "Europe's border-free zone expands" (http://news.bbc.co.uk/1/hi/7153490.stm). BBC News. 21 December 2007. . Retrieved 28 July 2011.

[47] "Poland.pl – White Stork – About White Stork" (http://storks.poland.pl/about_stork/index.htm). Storks.poland.pl. . Retrieved 2009-05-06.

[48] (Polish) "Wybrzeże Morza Bałtyckiego" (http://zalewszczecinski.net/akweny/wybrzeze-morza-baltyckiego). *www.zalewszczecinski.net*. . Retrieved 2009-11-16.

[49] "News Poland | Minorities make the news" (http://www.newpolandexpress.pl/polish_news_story-3634-minorities_make_the_news_. php). Newpolandexpress.pl. 2011-10-15. . Retrieved 2011-11-02.

[50] "Transgender woman poised for seat in Poland's new parliament" (http://www.telegraph.co.uk/news/worldnews/europe/poland/ 8817279/Transgender-woman-poised-for-seat-in-Polands-new-parliament.html). London: Telegraph. 2011-10-10. . Retrieved 2011-11-02.

[51] Davies, Norman (1996). *Europe: A History* (http://books.google.com/books?id=jrVW9W9eiYMC&pg=PA699). Oxford University Press. p. 699. ISBN 0198201710. .

[52] Grushenko, Kateryna (12 November 2010). "Polish representative: 'Poland is ready to help Ukraine as long as you are interested'" (http:// www.kyivpost.com/news/guide/world-in-uktaine/detail/89674/). Kyiv Post. . Retrieved 28 July 2011.

[53] "Bordering on madness: Belarus mistreats its Polish minority" (http://economist.com/displaystory.cfm?story_id=4085710). The Economist. 16 June 2005. . Retrieved 28 July 2011.

[54] "Jerzy Buzek elected President of the European Parliament" (http://www.europarl.europa.eu/sides/getDoc.do?type=IM-PRESS& reference=20090713IPR58063&language=EN). European Parlament. 2009-07-14. . Retrieved 2010-12-12.

[55] (Polish) "Strategia Bezpieczeństwa Narodowego RP" (http://www.wp.mil.pl/pliki/File/zalaczniki_do_stron/SBN_RP.pdf) (PDF). *www.wp.mil.pl*. . Retrieved 2008-09-26.

[56] Day, Matthew (5 August 2008). "Poland ends army conscription" (http://www.telegraph.co.uk/news/worldnews/europe/poland/ 2505447/Poland-ends-army-conscription.html). London: Telegraph. . Retrieved 28 July 2011.

[57] "Polska zakończyła udział w misjach po auspicjami ONZ - Wiadomości z kraju i ze świata - Gazeta Prawna - Partner pracodawcy, narzędzie specjalisty" (http://www.gazetaprawna.pl/wiadomosci/artykuly/382706,polska_zakonczyla_udzial_w_misjach_po_auspicjami_onz.html). Gazetaprawna.pl. 2009-12-31. . Retrieved 2011-11-02.

[58] "Accident Database" (http://www.airdisaster.com/cgi-bin/view_details.cgi?date=04102010®=101&airline=Polish+Air+Force). AirDisaster.com. . Retrieved 2010-12-12.

[59] "Senior Polish figures killed in plane crash" (http://news.bbc.co.uk/2/hi/europe/8613395.stm). BBC. 2010-04-11. .

[60] "– 15 tys. zimowych mundurów trafi do jednostek" (http://www.policja.pl/portal/pol/1/59960/ 15_tys_zimowych_mundurow_trafi_do_jednostek.html). Policja.pl. 30 September 2009. . Retrieved 28 July 2011.

[61] "Nowe radiowozy dla policji" (http://moto.onet.pl/1545847,1,nowe-radiowozy-dla-policji,artykul.html?node=2). Moto.onet.pl. 3 March 2009. . Retrieved 28 July 2011.

[62] "Country and Lending Groups | Data" (http://data.worldbank.org/about/country-classifications/ country-and-lending-groups#High_income). Data.worldbank.org. . Retrieved 2010-11-09.

[63] Pleitgen, Fred; Davies, Catriona (29 Jun 2010). "How Poland became only EU nation to avoid recession" (http://articles.cnn.com/ 2010-06-29/world/poland.economy.recession_1_poland-transition-economies-eastern-europe). *CNN*. . Retrieved 12 October 2011.

[64] "Central Europe Risks Downgrades on Worsening Finances (Update1)" (http://www.bloomberg.com/apps/news?pid=newsarchive& sid=apz_BEuHrNpI). Bloomberg.com. 2009-09-21. . Retrieved 2010-01-27.

[65] "Zloty to Gain, Says LBBW, Most Accurate Forecaster (Update1)" (http://www.bloomberg.com/apps/news?pid=newsarchive& sid=a7k1g_PkuROs#). Bloomberg.com. 2009-10-09. . Retrieved 2010-01-27.

[66] "Banking Sector in Poland | Thomas White International" (http://www.thomaswhite.com/explore-the-world/emerging-market-spotlight/ poland-banking.aspx). Thomaswhite.com. 2011-09-16. .

[67] " Poland in the Lead (http://www.warsawvoice.pl/archiwum.phtml/5473/)", *The Warsaw Voice*, September 2002. Retrieved on August 11, 2007.

[68] "Polish Information and Foreign Investment Agency. News" (http://www.paiz.gov.pl/nowosci/?id_news=1297&lang_id=12). *www.paiz.gov.pl*. . Retrieved 2008-09-26.

[69] "GDP per capita in PPS" (http://epp.eurostat.ec.europa.eu/tgm/table.do?tab=table&init=1&plugin=1&language=en& pcode=tsieb010). Eurostat. . Retrieved 2010-06-25.

[70] Bloomberg Businessweek. 2 December 2011. Official: Poland to be ready for euro in 4 years (http://www.businessweek.com/ap/ financialnews/D9RCGVPG2.htm)

[71] Schwab, Klaus. "The Global Competitiveness Report 2010-2011" (http://www3.weforum.org/docs/ WEF_GlobalCompetitivenessReport_2010-11.pdf). World Economic Forum. . Retrieved 25 April 2011.

[72] Shapiro, Robert J.. "Foreign Direct Investments in Developing Nations: Issues in Telecommunications and the Modernization of Poland" (http://www.investmentsecurity.org/wp-content/uploads/2011/02/CEIS-FDI-Report_Apr-2011.pdf). CEIS. . Retrieved 27 April 2011.

[73] "Waking up to the new economy: Ernst & Young's 2010 European attractiveness survey" (http://www.ey.com/Publication/vwLUAssets/Attractiveness_survey_2010_EU/$FILE/Attractiveness_survey_2010_EU.pdf). Ernst & Young. . Retrieved 25 April 2011.

[74] "Communication on the average monthly salary in enterprise sector excluding payments from profit awards in December 2010" (http://finanse.wp.pl/kat,58434,title,GUS-podal-dane-o-wynagrodzeniach,wid,13051878,wiadomosc.html?ticaid=1ba11) (in (Polish)). Stat.gov.pl. . Retrieved 2010-11-04.

[75] "OECD Economic Outlook No. 82 – Poland" (http://web.archive.org/web/20070328163217/http://www.oecd.org/dataoecd/6/32/20213254.pdf) (PDF). Archived from the original (http://www.oecd.org/dataoecd/6/32/20213254.pdf) on 2007-03-28. . Retrieved 2010-04-12.

[76] "Statistic Office of Poland(GUS)" (http://www.stat.gov.pl/bdr_n/app/wybrane_cechy.wymiary) (in (Polish)). Stat.gov.pl. . Retrieved 2010-11-19.

[77] "Poles getting rich quick..." (http://www.thenews.pl/business/artykul141422_poles-getting-rich-quick-.html). Polskie Radio, thenews.pl. . Retrieved 2010-10-13.

[78] Jędrzej Bielecki. "Polacy są w światowej czołówce bogacących się narodów" (http://forsal.pl/artykuly/458053,polacy_sa_w_swiatowej_czolowce_bogacacych_sie_narodow.html). Dziennik Gazeta Prawna. . Retrieved 2010-10-13.

[79] "Coraz więcej Azjatów emigruje do Polski" (http://fakty.interia.pl/prasa/news/coraz-wiecej-azjatow-emigruje-do-polski,1326394,16). Interia.pl. . Retrieved 2010-10-13.

[80] "Eurostat February 2008 – Euro area unemployment stable at 7.1%" (http://web.archive.org/web/20080726044250/http://epp.eurostat.ec.europa.eu/pls/portal/docs/PAGE/PGP_PRD_CAT_PREREL/PGE_CAT_PREREL_YEAR_2008/PGE_CAT_PREREL_YEAR_2008_MONTH_04/3-01042008-EN-AP.PDF). Web.archive.org. 26 July 2008. Archived from the original (http://epp.eurostat.ec.europa.eu/pls/portal/docs/PAGE/PGP_PRD_CAT_PREREL/PGE_CAT_PREREL_YEAR_2008/PGE_CAT_PREREL_YEAR_2008_MONTH_04/3-01042008-EN-AP.PDF) on 2008-07-26. . Retrieved 28 July 2011.

[81] "Polish economy seen as stable and competitive" (http://www.wbj.pl/article-51029-polish-economy-seen-as-stable-and-competitive.html). Warsaw Business Journal. 9 September 2010. . Retrieved 28 July 2011.

[82] "How Poland became only EU nation to avoid recession" (http://articles.cnn.com/2010-06-29/world/poland.economy.recession_1_poland-transition-economies-eastern-europe?_s=PM:WORLD). CNN. 29 June 2010. . Retrieved 28 July 2011.

[83] ATLAS AMONGST THE BEST POLISH BRANDS (http://dziennik.dziennikbudowy.pl/ang.nsf/0/2CE31E39A26A64F3C1256C31002C1FA4?OpenDocument). Retrieved 21 November 2010.

[84] "fDi: Poland Primed for Golden Decade" (http://www.ginannebrownell.com/2010/10/fdi-poland-primed-for-golden-decade/). GinanneBrownell.com. 8 October 2010. . Retrieved 28 July 2011.

[85] (Polish) "Lista 500 największych polskich firm" (http://www.lista500.polityka.pl/rankings/show). *www.lista500.polityka.pl*. . Retrieved 2011-08-27.

[86] (English) "Travel And Tourism in Poland" (http://www.euromonitor.com/Travel_And_Tourism_in_Poland). *www.euromonitor.com*. . Retrieved 2009-10-12.

[87] (English) "UNTWO World Tourism Barometer, Vol.5 No.2" (http://www.tourismroi.com/Content_Attachments/27670/File_633513750035785076.pdf). *www.tourismroi.com*. . Retrieved 2009-10-12.

[88] "Use of renewable fuel in Central and Eastern Europe" (http://www.frost.com/prod/servlet/cpo/205103205.htm). Frost.com. 1 March 2009. . Retrieved 28 July 2011.

[89] "EU Commission – Energy factsheet P74" (http://ec.europa.eu/energy/publications/statistics/doc/2010_energy_transport_figures.pdf) (PDF). . Retrieved 2011-07-28.

[90] "National Road Rebuilding Program (Polish)" (http://bip.mswia.gov.pl/portal/bip/175/17721/Narodowy_Program_Przebudowy_Drog.html). Bip.mswia.gov.pl. 2006-02-16. . Retrieved 2011-07-28.

[91] "Super pociągi zamiast autostrad (Polish)" (http://www.tvn24.pl/12690,1634983,0,1,super-pociagi-zamiast-autostrad,wiadomosc.html). *TVN24*. December 23, 2009. . Retrieved December 25, 2009.

[92] (English) Richard Francis Mould (1993). *A century of X-rays and radioactivity in medicine: with emphasis on photographic records of the early years* (http://books.google.com/?id=IXPz7bVR7g0C&printsec=frontcover&dq=A+century+of+x-rays+and+radioactivity+in+medicine:&q=). p. 19. ISBN 978-0-7503-0224-1. .

[93] Newswire Poland Emerges as the European R&D Hub Despite Favorable Conditions in Asia Pacific (http://www.newswiretoday.com/news/26490/)

[94] Polish Information and Foreign Investment Agency Poland – R&D centre (http://web.archive.org/web/20080219145055/http://www.paiz.gov.pl/index/?id=7b7a53e239400a13bd6be6c91c4f6c4e)

[95] (English) KPMG Sp. z o.o.. "Why Poland?" (http://www.paiz.gov.pl/files/?id_plik=7513). *www.paiz.gov.pl*. p. 3. . Retrieved 2011-08-27. "Over 80% of foreign investors see the results of their investments to date as positive or very positive and none of the studied companies reported a negative opinion."

[96] (English) "Key data on IT and telecoms market in Poland, 2004–2006" (http://web.archive.org/web/20061108212651/http://www.sat.org.au/reviews/articles_pl_middle_ages.htm). *www.itandtelecompoland.com*. Archived from the original (http://www.itandtelecompoland.com/index.php?item=2) on 2006-11-08. . Retrieved 2008-09-24.

[97] *The Expulsion of 'German' Communities from Eastern Europe at the end of the Second World War* (http://cadmus.iue.it/dspace/
      bitstream/1814/2599/1/HEC04-01.pdf), Steffen Prauser and Arfon Rees, European University Institute, Florense. HEC No. 2004/1. p.29

[98] (English) Michał Buchowski, Katarzyna Chlewińska. "Tolerance and Cultural Diversity Discourses in Poland" (http://www.eui.eu/
      Projects/ACCEPT/Documents/Research/wp1/ACCEPTPLURALISMWp1BackgroundreportPoland.pdf). *www.eui.eu.* . Retrieved
      2011-08-27.

[99] (English) Jan Repa (2007-01-05). "Poles return to Russian language" (http://news.bbc.co.uk/2/hi/europe/6233821.stm).
      news.bbc.co.uk. . Retrieved 2011-08-27. "In former satellite countries like Hungary or Poland, knowledge of Russian dwindled rapidly — to be
      replaced by English and German."

[100] Doughty, Steve (2006-04-25). "UK lets in more Poles than there are in Warsaw" (http://www.dailymail.co.uk/news/article-384121/
      UK-lets-Poles-Warsaw.html). London: Dailymail.co.uk. . Retrieved 2010-04-12.

[101] (English) Alexi Mostrous, Christine Seib (February 16, 2008). "Tide turns as Poles end great migration" (http://www.timesonline.co.
      uk/tol/news/uk/article3378877.ece). *www.timesonline.co.uk* (London). . Retrieved 2011-08-27. "The Times has established that, for the first
      time since they began arriving en masse four years ago, more UK-based Poles are returning to their homeland than are entering Britain."

[102] "Polish Diaspora (Polonia) Worldwide" (http://culture.polishsite.us/articles/art79fr.htm). Culture.polishsite.us. . Retrieved 2010-04-12.

[103] "Centers of Polish Immigration in the World — USA and Germany" (http://culture.polishsite.us/articles/art90fr.htm).
      Culture.polishsite.us. 2003-03-15. . Retrieved 2010-04-12.

[104] "Maly Rocznik Statystyczny Polski 2009" (http://www.stat.gov.pl/cps/rde/xbcr/gus/PUBL_oz_maly_rocznik_statystyczny_2009.
      pdf) (in (Polish)) (PDF). . Retrieved 2009-09-26.

[105] *"94% Polaków wierzy w Boga"* (http://www.ekumenizm.pl/content/article/20080925183042429.htm). Ekumenizm.pl. 2008-09-25. .
      Retrieved 2010-04-12.

[106] Weekly Mass Attendance of Catholics in Nations with Large Catholic Populations, 1980–2008 (http://web.archive.org/web/
      20080214110918/http://cara.georgetown.edu/bulletin/international.htm) — World Values Survey (WVS)

[107] (Polish) Centrum Badania Opinii Społecznej (*Centre for Public Opinion Research (Poland)* CBOS). Komunikat z badań; Warszawa,
      Marzec 2005. Co łączy Polaków z parafią? (http://www.cbos.pl/SPISKOM.POL/2005/K_049_05.PDF) Preface. Retrieved 2007-12-14.

[108] Wilde, Robert. "Pope John Paul II 1920–2005" (http://europeanhistory.about.com/od/religionandthought/a/biojohnpaulii.htm).
      About.com. . Retrieved 2009-01-01.

[109] Domínguez, Juan: 2005

[110] "Pope John Paul II and Communism" (http://www.religion-cults.com/pope/communism.htm). Public domain text. May be distributed
      freely. No rights reserved.. . Retrieved 2009-01-01.

[111] (Polish) Dr Zbigniew Pasek, Jagiellonian University, "Wyznania religijne" (http://web.archive.org/web/20061128165649/http://
      www.religioznawstwo.uj.edu.pl/syllabusy/pasek-wrwp.rtf). Archived from the original (http://www.religioznawstwo.uj.edu.pl/
      syllabusy/pasek-wrwp.rtf) on November 28, 2006. . Retrieved 2007-09-15. Further reading: Ustawa o gwarancjach wolności sumienia i
      wyznania z dnia 17 V 1989 z najnowszymi nowelizacjami z 1997 roku.

[112] (Polish) Michał Tymiński, "Kościół Zielonoświątkowy" (http://web.archive.org/web/20050102151031/http://www.kz.pl/index.
      php?p=13&id=3&i=8). Archived from the original (http://www.kz.pl/index.php?p=13&id=3&i=8) on January 2, 2005. . Retrieved
      2007-09-14.

[113] (Polish) Dr. Paweł Borecki, "Opinia prawna dotycząca religii w szkole" (http://www.racjonalista.pl/kk.php/s,5534). Kateda Prawa
      Wyznaniowego Uniwersytetu Warszawskiego. . Retrieved 2007-09-14.

[114] (Polish) Wirtualna Polska, Wiadomości. "Polacy przeciwni wliczaniu ocen z religii do średniej" (http://wiadomosci.wp.pl/
      kat,9911,wid,9125933,wiadomosc.html?ticaid=1478d). . Retrieved 2007-09-14.

[115] (Polish) Olga Szpunar, "Dorośli chcą religii w szkole" (http://miasta.gazeta.pl/krakow/1,35798,4360977.html). Gazeta Wyborcza
      Kraków. . Retrieved 2007-09-15.

[116] "Poland Guide: The Polish health care system, An introduction: Poland's health care is based on a general" (http://www.justlanded.com/
      english/Poland/Poland-Guide/Health/The-Polish-health-care-system). Justlanded.com. . Retrieved 2011-07-28.

[117] "Polish hospitals" (http://polandpoland.com/polish_hospitals.html). Polandpoland.com. . Retrieved 2011-07-28.

[118] "The Quality of Medical Treatment and Surgery in Poland" (http://www.articlesbase.com/medicine-articles/
      the-quality-of-medical-treatment-and-surgery-in-poland-602240.html). Articlesbase.com. . Retrieved 2011-07-28.

[119] "WHO | Poland" (http://www.who.int/countries/pol/en/). Who.int. 2011-05-17. . Retrieved 2011-07-28.

[120] OECD (2009). "The impact of the 1999 education reform in Poland" (http://www.pisa.oecd.org/dataoecd/50/26/45721631.doc). .
      Retrieved 2010-09-17.

[121] (English) "Range of rank on PISA 2006 science scale" (http://www.oecd.org/dataoecd/42/8/39700724.pdf) (PDF). *www.oecd.org.* .
      Retrieved 2008-09-24.

[122] Adam Zamoyski, The Polish Way: A Thousand Year History of the Poles and Their Culture (http://www.amazon.co.uk/
      Polish-Way-Thousand-History-Culture/dp/978-0-7818-0200-0). Published 1993, Hippocrene Books, Poland, ISBN 978-0-7818-0200-0

[123] Ministry of Foreign Affairs of Poland, 2002–2007, AN OVERVIEW OF POLISH CULTURE. (http://www.poland.gov.pl/Culture,484.
      html) Access date 12-13-2007.

[124] (English) "Nicolaus Copernicus" (http://www.britannica.com/EBchecked/topic/136591/Nicolaus-Copernicus). *www.britannica.com.*
      . Retrieved 2008-10-10.

[125] (French) Rey Alain (1993). *Le petit Robert 2 : (dictionnaire universel des noms propres, alphabétique et analogique)*. INIST-CNRS, Cote INIST : L 22712: Le Robert, Paris, FRANCE. ISBN 978-2-85036-210-1.

[126] (English) Michael Kennedy, ed (2004). *The Concise Oxford dictionary of music*. Oxford University Press. ISBN 978-0-19-860884-4. p. 141

[127] (French) "Maria Sklodowska. La jeunesse" (http://mariecurie.science.gouv.fr/portrait/portrait1_1.php). *mariecurie.science.gouv.fr*. . Retrieved 2008-10-10.

[128] (Polish) Koca, B. (2006). "Polish Literature – The Middle Ages (Religious writings)" (http://web.archive.org/web/20061108212651/ http://www.sat.org.au/reviews/articles_pl_middle_ages.htm). Archived from the original (http://www.sat.org.au/reviews/ articles_pl_middle_ages.htm) on 2006-11-08. . Retrieved 10 December 2006.

[129] (English) Zdzislaw Najder (1998). "Profiles – Joseph Conrad" (http://www.culture.pl/en/culture/artykuly/os_conrad_joseph). *www.culture.pl*. . Retrieved 2008-09-30.

[130] http://www.coe.int/t/dg3/children/corporalpunishment/pdf/EnglishQuestionAnswer_en.pdf

[131] "Dr. Sławomir Łodziński, Helsinki Foundation for Human Rights, "The Protection of National Minorities in Poland"" (http://www. minelres.lv/reports/poland/poland_NGO.htm). Minelres.lv. . Retrieved 2011-07-28.

[132] (English) "The Music Courts of the Polish Vasas" (http://www.semper.pl/muzyczne_dwory_summary.pdf). *www.semper.pl*. p. 244. . Retrieved 2009-05-13.

[133] Busfield, Steve (2006-02-21). "Guardian wins design award" (http://media.guardian.co.uk/site/story/0,,1714643,00.html). *The Guardian* (London). .

[134] "Facts on the Nobel Prize in Literature" (http://nobelprize.org/nobel_prizes/literature/shortfacts.html). Nobelprize.org. 2009-10-05. . Retrieved 2011-07-28.

[135] Adam Gopnik (June 5, 2007). "Szymborska's 'View': Small Truths Sharply Etched" (http://www.npr.org/templates/story/story. php?storyId=10721773). npr.org. . Retrieved December 12, 2010.

[136] (English) "Szydłowiec" (http://web.archive.org/web/20060623020259/http://www.szydlowiec.pl/grafika/index/szydl1.pdf). *www.szydlowiec.pl*. p. 9. Archived from the original (http://www.szydlowiec.pl/grafika/index/szydl1.pdf) on June 23, 2006. . Retrieved 2009-04-23.

[137] Many designs imitated the arcaded courtyard and arched loggias of the Wawel palace. (English) Michael J. Mikoś. "RENAISSANCE CULTURAL BACKGROUND" (http://www.staropolska.pl/ang/renaissance/Mikos_renaissance/Cultural_r.html). *www.staropolska.pl*. p. 9. . Retrieved 2009-04-23.

[138] (English) John Stanley (March–June 2004). "Literary Activities and Attitudes in the Stanislavian Age in Poland (1764–1795): A Social System?" (http://findarticles.com/p/articles/mi_qa3763/is_200403/ai_n9363971/?tag=content;col1). *findarticles.com*. . Retrieved 2009-04-23.

[139] "FIFA World Cup Statistics-Poland" (http://www.fifa.com/worldfootball/statisticsandrecords/associations/association=pol/worldcup/ index.html). FIFA. . Retrieved December 12, 2010.

[140] "FIFA Statistics – Poland" (http://www.fifa.com/worldfootball/statisticsandrecords/associations/association=pol/othertournaments/ index.html). . Retrieved December 12, 2010.

[141] "Poland hosts Euro 2012!" (http://www.warsaw-life.com/poland/euro-2012). warsaw-life.com. . Retrieved December 12, 2010.

[142] "Speedway World Cup: Poland win 2010 Speedway World Cup" (http://www.worldspeedway.com/artman/publish/article_13423. shtml). worldspeedway.com. . Retrieved December 18, 2010.

[143] Encyklopedia kuchni: "Kuchnia polska." (http://www.kuchnia.tv/kuchnia_encyklopedia-kuchni_185_kuchnia-polska.html) Kuchnia.tv, ul. W. Sikorskiego 902-758 Warszawa, 2011.

[144] http://news.bbc.co.uk/2/hi/uk/6359363.stm

[145] Steven J. Zaloga; Richard Hook (21 January 1982). *The Polish Army 1939–45* (http://books.google.com/books?id=AAdYFeW2fnoC& pg=PA3). Osprey Publishing. pp. 3–. ISBN 9780850454178. . Retrieved 6 March 2011.

[146] Jerzy Jan Lerski (1996). *Historical dictionary of Poland, 966–1945* (http://books.google.com/books?id=QTUTqE2difgC&pg=PA18). Greenwood Publishing Group. pp. 18–. ISBN 9780313260070. . Retrieved 6 March 2011.

[147] E. Garrison Walters (1988). *The other Europe: Eastern Europe to 1945* (http://books.google.com/books?id=64VpSBd7xUcC& pg=PA276). Syracuse University Press. pp. 276–. ISBN 9780815624400. . Retrieved 6 March 2011.

[148] Norman Davies, *God's Playground: A History of Poland*, Columbia University Press, 2005, ISBN 0-231-12819-3, Google Print p.344 (http://books.google.com/books?id=EBpghdZeIwAC&pg=PA344&dq="Armia+Krajowa"+largest& ei=hTrMR_W-G4mWzASVs9GtCQ&sig=iE7xbtRu3rvEsVZZgCeUsqEqj6s)

[149] Gregor Dallas, *1945: The War That Never Ended*, Yale University Press, 2005, ISBN 0-300-10980-6, Google Print, p.79 (http://books. google.com/books?id=LXdVF6LmTa8C&pg=PA79&dq="Armia+Krajowa"+largest&as_brr=3&ei=RjvMR6KnPJPAzAT-ppWvCQ& sig=Ksba8pTs5pu55YiAqseCLy6Kl5k)

[150] Mark Wyman, *DPs: Europe's Displaced Persons, 1945–1951*, Cornell University Press, 1998, ISBN 0-8014-8542-8, Google Print, p.34 (http://books.google.com/books?id=lHNw7MnsmlYC&pg=PA34&dq="Armia+Krajowa"+largest&lr=&as_brr=3& ei=NzzMR_mOIJGSzQSb7cSwCQ&sig=kv3oN5z3YgAgcT8Vgy4aIFRHknE)

[151] See for example: Leonid D. Grenkevich in The Soviet Partisan Movement, 1941–44: A Critical Historiographical Analysis, p.229 or Walter Laqueur in The Guerilla Reader: A Historical Anthology, (New York, Charles Scribiner, 1990, p.233.

[152] Velimir Vukšić (23 July 2003). *Tito's partisans 1941–45* (http://books.google.com/books?id=SLix5hc4WRgC&pg=PA11). Osprey
       Publishing. pp. 11–. ISBN 9781841766751. . Retrieved 1 March 2011.

[153] Anna M. Cienciala, THE COMING OF THE WAR AND EASTERN EUROPE IN WORLD WAR II. (http://web.ku.edu/~eceurope/
       hist557/lect16.htm), History 557 Lecture Notes

External links

Movie (on-line)

- Animated history of Poland, (PARP, Expo 2010 Shanghai China) (http://www.youtube.com/
  watch?v=2DrXgj1NwN8)
- Borders of Poland, A.D. 990–2008 (http://www.youtube.com/watch?v=eAVVWlUywO0)
- Poland under German occupation 1939–1945 (http://www.youtube.com/watch?v=nSzejCWXKhA)

General

- Poland.gov.pl – Polish national portal (http://www.poland.gov.pl/)
- Polish Information (http://www.paiz.gov.pl/index/?lang_id=1)
- Ministry of Foreign Affairs (http://www.msz.gov.pl/)
- Chief of State and Cabinet Members (https://www.cia.gov/library/publications/world-leaders-1/
  world-leaders-p/poland.html)
- Poland (https://www.cia.gov/library/publications/the-world-factbook/geos/pl.html) entry at *The World
  Factbook*
- Poland (http://ucblibraries.colorado.edu/govpubs/for/poland.htm) at *UCB Libraries GovPubs*
- History of Ancient Poland (http://www.theancientweb.com/explore/content.aspx?content_id=24)
- Poland (http://www.dmoz.org/Regional/Europe/Poland/) at the Open Directory Project
- Wikimedia Atlas of Poland

Culture

- Commonwealth of Diverse Cultures: Poland's Heritage (http://www.commonwealth.pl/)
- News about the Polish culture (http://www.culture.pl/en/culture/)

Travel

- Poland travel guide from Wikitravel
- Poland's Official Travel Website (http://www.poland.travel/en-us/)
- About Poland (http://www.international-travel-tours.com/poland/about-poland.html)
- Photos from Poland (http://www.klub-beskid.com/album-pologne/index.php?lang=english)

kbd:Полшэ gag:Polşa mrj:Польша ltg:Puoleja xmf:პოლონეთი koi:Польска rue:Польско

Bylin

<table>
<tr><td colspan="2" align="center">Bylin</td></tr>
<tr><td colspan="2" align="center">— Village —</td></tr>
<tr><td colspan="2" align="center">Bylin</td></tr>
<tr><td colspan="2" align="center">Coordinates: 52°20′N 17°8′E</td></tr>
<tr><td>Country</td><td>Poland</td></tr>
<tr><td>Voivodeship</td><td>Greater Poland</td></tr>
<tr><td>County</td><td>Poznań County</td></tr>
<tr><td>Gmina</td><td>Kleszczewo</td></tr>
</table>

Bylin [ˈbɨlin] is a village in the administrative district of Gmina Kleszczewo, within Poznań County, Greater Poland Voivodeship, in west-central Poland.[1] It lies approximately 17 km (11 mi) south-east of the regional capital Poznań.

References

[1] "Central Statistical Office (GUS) - TERYT (National Register of Territorial Land Apportionment Journal)" (http://www.stat.gov.pl/ broker/access/prefile/listPreFiles.jspa) (in Polish). 2008-06-01. .

Main_Page

<table>
<tr><td rowspan="2"><h2>Welcome to Wikipedia,</h2>
the free encyclopedia that anyone can edit.
0 articles in English</td><td>• Arts
• Biography
• Geography</td><td>• History
• Mathematics
• Science</td><td>• Society
• Technology
• **All portals**</td></tr>
</table>

From Wikipedia's newest content:

- ... that **Mojo Mathers** *(pictured)* became New Zealand's first deaf MP during the 2011 general election?
- ... that before Brighton's **Montpelier** suburb developed, three people lived on the hilly site – including an eccentric corporal who lived in a cave and fired celebratory pistols on military anniversaries?
- ... that after giving seven days of speeches before 500 Hindu scholars, **Jagadguru Kripalu Maharaj** was named fifth Jagadguru (world teacher) at the age of 34?
- ... that female baseball player **Mary Rountree** earned a medical degree by studying in the off-season?
- ... that *Three Sisters*, a musical by Oscar Hammerstein II and Jerome Kern, played for only two months in London in 1934, and didn't debut in America until 2011?
- ... that 30-to-120-centimetre (12 to 47 in) tall **hairy pagodas** can be found throughout eastern North America?

[[File:Bolotnaya-wiki-2.jpg|100x100px

alt=Protests in Bolotnaya Square, Moscow|link=File:Bolotnaya-wiki-2.jpg

- At least five people are killed and more than one hundred are wounded in **a murder–suicide attack** in Liège, Belgium.
- The Constituent Assembly of Tunisia elects **Moncef Marzouki** as the President of Tunisia.
- The **United Nations Climate Change Conference** in Durban, South Africa, closes with an agreement to establish a new treaty to limit carbon emissions.
- Allegations of flaws in the Russian legislative elections trigger the country's **largest protests** *(pictured)* since the dissolution of the Soviet Union.
- Iran files a formal complaint to the UN Security Council regarding a U.S. RQ-170 Sentinel aircraft **seized** within its territory.

December 14: Day of the Children Killed During the War in Abkhazia (1992); **Day of the Martyred Intellectuals** in Bangladesh (1971); **Monkey Day**

- 557 – **A large earthquake** severely damaged the city of Constantinople.
- 1782 – In Avignon, France, the **Montgolfier brothers** conducted their first test of their hot air balloon.
- 1994 – Construction on the **Three Gorges Dam** *(pictured)* began on the Yangtze River in China.
- 1999 – **Torrential rains** caused flash floods in Vargas, Venezuela, resulting in tens of thousands of deaths, the destruction of thousands of homes, and the complete collapse of the state's infrastructure.
- 2009 – The **Tino Rangatiratanga flag** representing the Māori people was recognized officially by the government of New Zealand.

More anniversaries: December 13 – **December 14** – December 15

It is now December 14, 2011 (UTC) – Refresh this page [1]

(Check back later for today's.)

The **Three Sisters** is a sandstone rock formation in the Blue Mountains of New South Wales, Australia, near the town of Katoomba. It is a popular bushwalking destination and visitors may descend from the Three Sisters down into the Jamison Valley via a series of 800 steel and stone steps.

Photo: JJ Harrison

Recently featured: Brown-throated sloth – Anscombe's quartet – American Bird Grasshopper

Other areas of Wikipedia

- **Community portal** – Bulletin board, projects, resources and activities covering a wide range of Wikipedia areas.
- **Help desk** – Ask questions about using Wikipedia.
- **Local embassy** – For Wikipedia-related communication in languages other than English.
- **Reference desk** – Serving as virtual librarians, Wikipedia volunteers tackle your questions on a wide range of subjects.
- **Site news** – Announcements, updates, articles and press releases on Wikipedia and the Wikimedia Foundation.
- **Village pump** – For discussions about Wikipedia itself, including areas for technical issues and policies.

Wikipedia languages

This Wikipedia is written in English. Started in 2001, it currently contains 0 articles. Many other Wikipedias are available; some of the largest are listed below.

- More than 700,000 articles: [//.org/wiki/ de] · [//.org/wiki/ es] · [//.org/wiki/ fr] · [//.org/wiki/ it] · [//.org/wiki/ nl] · [//.org/wiki/ ja] · [//.org/wiki/ pl] · [//.org/wiki/ pt] · [//.org/wiki/ ru]
- More than 150,000 articles: [//.org/wiki/ ar] · [//.org/wiki/ id] · [//.org/wiki/ ca] · [//.org/wiki/ cs] · [//.org/wiki/ da] · [//.org/wiki/ eo] · [//.org/wiki/ fa] · [//.org/wiki/ ko] · [//.org/wiki/ hu] · [//.org/wiki/ no] · [//.org/wiki/ ro] · [//.org/wiki/ sr] · [//.org/wiki/ fi] · [//.org/wiki/ sv] · [//.org/wiki/ vi] · [//.org/wiki/ tr] · [//.org/wiki/ uk] · [//.org/wiki/ zh]
- More than 50,000 articles: [//.org/wiki/ ms] · [//.org/wiki/ bg] · [//.org/wiki/ et] · [//.org/wiki/ el] · [//.org/wiki/ simple] · [//.org/wiki/ eu] · [//.org/wiki/ gl] · [//.org/wiki/ he] · [//.org/wiki/ hr] · [//.org/wiki/ lt] · [//.org/wiki/ nn] · [//.org/wiki/ sk] · [//.org/wiki/ sl] · [//.org/wiki/ sh] · [//.org/wiki/ th]

[//meta.wikimedia.org/wiki/List_of_Wikipedias Complete list of Wikipedias]

References

[1] http://en.wikipedia.org/wiki/Main_page

Article Sources and Contributors

Poland *Source*: http://en.wikipedia.org/w/index.php?title=Poland *Contributors*: 130.94.122.xxx, 217.99.96.xxx, 2eet2eet, 334a, 3sen, 4 bity muzyki, 4darek, 9cds, A bit iffy, A-giau, A. (pl), A.Jalil, A2Kafir, A8UDI, ABCD, ARC Gritt, Aaron Brenneman, Abductive, Abeg92, Academic Challenger, Achaemenes II, Acroterion, Adam Rock, AdamJacobMuller, AdamantlyMike, Adambro, Adamcisme123, Adashiel, Adavis444, Adz, Ae-a, Afinati, Agnus, Agquarx, Ahoerstemeier, AirdishStraus, AirlineMan333, Aivazovsky, Ajh1492, Akamad, Akanemoto, Albanman, Alcatraz014, AlefZet, Aleph-4, Aleph4, Alex.muller, Alex1100, Ali@gwc.org.uk, AliveFreeHappy, Allstarecho, Almankib, Altenmann, Amakuru, Ammon86, Amoruso, Ancheta Wis, Andre Engels, Andy M. Wang, Andy Marchbanks, Andycjp, Andypl, AngelOfSadness, Angela, Anger22, Angr, Annarapo, Annonnimus, Anonymous editor, Anonymous from the 21st century, Another disinterested reader, Antakaa, Antandrus, Antique Rose, Antman, Aotearoa, AppleJordan, Appleseed, Aquintero82, Aranherunar, Architect JDF, Arcillaroja, Argentina 678, Ari x, ArielGold, Aries1975, Arjun01, Army1987, Armydepot, Art LaPella, Arthena, Arvis21, Arwel Parry, AshLOVESgnr, Asidemes, Astian, Ataltane, Atlant, Atomekk, Attilios, Atwardow, Audirs8, Auntof6, Ausir, Austrian, Avala, Avaya1, Avb, Avicennais, AxG, Axegod12, Axt, AzaToth, Azalero, Azar66, B-2StealthEditor, BGManofID, BLT000, BRG, Babajobu, Backspace, Balcer, Bardhylius, Baristarim, Barry Kent, Barryob, Bart133, Barticus88, Bartsimpson, Baseball Watcher, Bastian17, Battlecry, Bayerischermann, Bazonka, Bazza 7, Bbatsell, Bbulzibar, Beagel, Beamathan, Beaumont, Beefyt, Beegees, Beetstra, Bejnar, Beland, Ben-Velvel, Bencherlite, Benqish, Benwildeboer, Berkunt, BernardaAlba, Betacommand, Bftravel, Biala Gwiazda, Bigboydog1, Bill3000, Bjung, Bkell, Black Falcon, Blair P. Houghton, Blood Red Sandman, BlueMars, Blueshade, Bmicomp, BoBo1488, Bob marley 118, Bobblewik, Bobbybrown07, Bobo192, Bobtbuilder, Bocianski, Bogdan, Bolo1729, Boraczek, Borgarde, Borisblue, Bornkaren, BostonMA, Boud, Boydenmb, Bprsolt Qaoddz, Brabblebrex, Brandmeister (old), Brandon, Briaboru, Brian0918, Brion VIBBER, Brisvegas, BritishWatcher, Brudasek, Brunopioss, Brutannica, Buaidh, Bukubku, Bull Market, BurgererSF, Burschenschafter, Bushy moustache, C filev, C.Fred, CHJL, CIreland, CLW, CT Cooper, CWY2190, Cactus.man, Caerwine, Caius2ga, Caknuck, CalJW, CalicoCatLover, Caltas, CambridgeBayWeather, Can't sleep, clown will eat me, CanadianPride, CanisRufus, Canterbury Tail, Cantus, Caponer, CaptainFugu, Carbonix, Carlwev, CarolGray, Casliber, Caspian, Cassini83, Catherine-Jayne, Cautious, Celinakat, Celtmist, Century0, CesarB, ChKa, Chabuk, Challisrussia, CharlesMartel, Charlietemps, CharlotteWebb, Chase me ladies, I'm the Cavalry, Cheesefondue, Cheeseloop, Chensiyuan, Chevinki, Chris G, Chris cufc, Chris the speller, ChrisGriswold, Chrisch, Chrischrischrischris, Chrishmt0423, Chrislk02, Christian List, Christoph.mamok, Christopher Kraus, Chromaticity, Chudzik321, Chumchum7, Chunky Rice, CieloEstrellado, Cimmerian praetor, Civil Engineer III, Closedmouth, CloutierFan02, Cocoaguy, Colchicum, Colonel Mustard, CommonsDelinker, Conscious, Consequencefree, Constantine Gorov, Constantzeanu, Conversion script, Coolszach2, Cormic, Corpx, Corvettegrandsport, Count Iblis, Coveritus, Covervisit, Cptnono, Craftypants, Crazycomputers, Crystallina, Cs-wolves, Csiew88, Ctjf83, Cubus1984, Cunningham, Curps, Cusio, Cybergoth, Cybershore, Czas Na Zywiec, Czyrko, D, D6, DBaba, DJ Clayworth, DJPeugeot, DMJohnston, DO'Neil, DRoninLA, DSuser, DVD R W, DX-Kevo, Daddy1234, Daffy123, Damian Radu, Damiang1997, Danaeb, Daniel De Mol, Daniel Quinlan, Daniil naumoff, Danim, Danlock2, Danny, Dannycas, Danziger100, Dariusz Szwed, Darklord517, Darth Kalwejt, Darwinek, Dave.Dunford, Davenbelle, Davewild, David Kernow, David Liuzzo, David Parker, David Richtman, Davidkazuhiro, Dawidbernard, Dbachmann, Dbenbenn, DeadEyeArrow, Deanna Bardenfleth, Deathmetal Greg, Debresser, Deflective, Dekisugi, Delldot, Demiurge, Denkealsobin, Denniss, Der Golem, DerHexer, Derek Ross, DerekJeter, Deschain97, Desiphral, Deuar, Digitalme, Digwuren, Dillard421, Dimkoa, Dina, Diogenes, Director958, Discospinster, Distal24, DivineIntervention, Dj3500, Dmanning, Dockingman, Docta247, Docu, Dodo von den Bergen, Don Alessandro, Doopdoop, Dougy109, Downyourpants, DrSlony, Dragoman1, Draktorn, Dreadstar, Drwhorules, DubaiTerminator, Dudeimoncoke, Dust Filter, DuttyYo, Dyamantese, Dycedarg, Dysepsion, Dysprosia, E Pluribus Anthony, EMT1871, ERcheck, ESkog, EarthPerson, Eclecticology, Edderso, EddyVadim, Editor at Large, Edivorce, Edward saint-ivan, Edward321, Ehheh, Ejoty, Ejército Rojo 1950, Ek8, Ekuprusevicius, El C, El-042, Eldin, Electionworld, Elementjunky, Elipongo, Elliskev, Elmondo21st, Elonka, Emax, Emigrant85, Emil Kastberg, Emote, Enchanter, EncycloPetey, Encyclopedist, Enforcer Of Lulz, Entryshare, Enviroboy, Enzo Aquarius, Eob, Ericamick, Erik1980, Esperant, Espoo, Etams, Eu.stefan, EugeneZelenko, Eumolpo, Ev, Evercat, Everyking, Everytime, Evlekis, FF2010, Fabhcún, Fabian Hassler, Farosdaughter, Farquaadhnchmn, Fasach Nua, FashionNugget, FayssalF, Fbeauregard, Fbifriday, Femto, Filemon, Filpaul, Filthdafugout, Filyo79, FinnishDriver, FireGaamer1122, Fireaxe888, Firstorm, FlagUploader, Flamarande, Flatterworld, Flockmeal, Fnagaton, Fordan, Fordie123, Foregone conclusion, Formeruser-81, Formulax, Foundert, Fox Mccloud, Fram, Francvs, Freak in the bunnysuit, Freakyfridge, Frees, FreplySpang, Freyr, Frstr, Frymaster, FunPika, Funandtrvl, Funnyhat, Future Perfect at Sunrise, Fvincent, G. Campbell, Gabbe, Gadfium, Gagabrain, Gaius Cornelius, Galati, Galileo01, Gam0001, Garion96, Garraisgreat, Garzo, Gaz, Gazilion, Gdo01, Geeks123, Geeoharee, Gene Nygaard, Geneb1955, Generalboss3, Geniu, Geopolio, George The Dragon, Georgeslegloupier, Gggh, Gherald, Ghost321, GhostPirate, GhostWhoVotes, Gibble, Gilliam, Gimmetrow, GinaDana, Giving is good, as long as you're getting, Gnesener1900, Gobonobo, Godfinger, Gogo Dodo, GoingBatty, Golbez, Golem Unity, Goltz, Good Olfactory, Goodlittleangel, GorillazFanAdam, GraemeL, Graham87, Gran2, Grayshi, Green Giant, GregorB, Grendelkhan, Greyhood, Ground Zero, Grubber, Gryffindor, Grzechooo, Gscshoyru, Gsklee, Gtg204y, Guinness man, Gurch, Gustavo Szwedowski de Korwin, Gwernol, Gznorneplatz, Gzornenplatz, H, H3xx, H8erade, HJ Mitchell, Hadal, Hagerman, Halibutt, Happyme22, Hardman1, Harmattan, Hatch68, Hayden120, Hdt83, Hebereke, Heimstern, Hello32020, Henry Flower, Hephaestos, Herr Kriss, Hillel, HistoriaPolska, Historian932, HistoricalAce, Hmains, Holod, Homunculus 2, Hoschimobi, Hottentot, Hu12, Husond, Hut 8.5, Ian Pitchford, Ianhowlett, Iapain wiki, Icehawk1223, Icollins68, Icseaturtles, Ief, Ignacy nowopolski, Ignatzmice, Ikiroid, Iksisy, Illythr, Imagod777, Imlepid, InShaneee, Indisciplined, Indon, Infogengachi, Informationguy, Insanely Beautiful, InsanityJ, Intangible2.0, Inter, Interfector, Internet Ignoramus, Iohannes Animosus, Irek Biernat, Irishguy, Istros, Istvan, Itai, Italiano111, Ivan89, J.delanoy, J04n, JLaTondre, JRWalko, Jacek Kendysz, Jacoplane, Jacurek, Jadog9495, Jagoperson, Jahanas, JamesBWatson, Jameswilson, Jankaspar, Janmariamazurczak, Janneman, Janus Shadowsong, JanusTehPheo, Jaranda, Jared Preston, Jaro7788, Jauerback, Jcaragonv, Jcc7386, Jcrook1987, JdeJ, Jecar, Jeff G., Jeff8765, Jeffhoy, Jeffmaylortx, Jensboot, Jerem43, Jerewak, Jh51681, Jhendin, Jiang, Jihg, Jiminy Krikkitt, Jimmyvanthach, Jizg, Jjbaruel, Jkelly, Jklin, Jkw0010, Jlujan69, Jni, JoeBlogsDord, Joffeloff, John, John K, John Smith M.D., Ph.D, John Vandenberg, Johnkenyon, Joho104, Jon513, Jonas Taylor, Jor, Jorunn, Joseph Solis in Australia, Josephschutz, Josiah Rowe, Jossi, Joy, Jpeob, Jpgordon, Jseidel, Jsport, Jtbelliott, Jukes85, Julekmen, Julo, Jumbuck, Jura2102, Juraj cisarik, Jurand, Jusjih, Jwissick, KConWiki, KJohansson, KOP, Kaihsu, Kajzderski, Kamil m, Kantokano, KapilTagore, Karasek, Karenjc, Karl, Karol Langner, Kaszkawal, Katarzyna, Kbrooks, Keegan, Keelm, KeithB, Kelisi, Kelly Martin, Kemiv, Khazar, Khoikhoi, King Bigos, Kinneyboy90, Kintetsubuffalo, Kirk diamond, Kman665, Kmcdm, Know Your NME, KnowledgeOfSelf, Knowledgebycoop, Knut, Knverma, Kobold 86, Kochamanita, KokkaShinto, Kollie1, KoloszRodos, Konkursor, Konrii, Konstable, Kopcik, Kotasik, Kotjze, Kotniski, Koyaanis Qatsi, Kozuch, Kpalion, Kpjas, KrakatoaKatie, Krautsaredummies, Kresen, Krich, Krupo, Krystiandl, Krzyzowiec, Ksenon, Kuba1000, Kubek15, Kubigula, Kubster, Kukini, Kungfuadam, Kuracpm1, Kuru, Kusma, Kwamikagami, Kwpolska, Kwsn, Kyokpae, L95slovenia, LA2, LAX, LMB, LOL, LOTkid, LUCPOL, Laflin22, Laforgue, Lambrook, Lampak, Latka, Laudaka, Lcamtuf, Le Fou, Leafyplant, Lear 21, Lebanonman19, Lectonar, LeeHunter, Lemonade100, Lenin and McCarthy, LeonWhite, Levineps, Leyo, Liam Plested, LibLord, LifeStar, Liftarn, Lightdarkness, Lightmouse, Lights, Likeminas, Ling.Nut, Listowy, Llewelyn MT, Llort, LlywelynII, Lockesdonkey, Logan, Logologist, Lolster, Lomisss, LonelyMarble, Looxix, Lord Voldemort, Loren36, Lost tourist, LoveToWiki, Lovesthespooge, Lrschum, Lst27, Luapnampahc, Luca Italy, Lucasm, Ludvikus, LugPaj, Luna Santin, Lynntoniolondon, Lysy, M.C., MD, MER-C, MJCdetroit, MJM74, MMich, MSGJ, MaLu, MackSalmon, Madriversnowboarder, Maelnuneb, Magister Mathematicae, Mailer diablo, Majorly, Maju wiki, Makemi, Makgraf, MakmoudHassan, Malcolm Farmer, Malcomosx, Malick78, Malik Shabazz, Malo, Mamaberry11, Man vyi, Marcin Suwalczan, Marcin.sobczyk, Marco polo, Marek69, MarekF, Mareklug, Mark Ryan, MarkGallagher, Markotek, Martarius, Marthainky, Martim33, Martinwilke1980, Marx01, Marzoregulus@hotmail.com, Marzoregulus@yahoo.co.ph, Master of Puppets, Mathiasrex, Matman from Lublin, Matt 989, Mattbr, Matthead, Mauricio Maluff, Mav, Maxamegalon2000, Maxumum12, MałyG, Mb nl, Mc6415, Mc6809e, McDogm, Meekywiki, Mega Prime, Meika, Melizabeth, Meow kitties, Merbabu, Mercury, Mercurywoodrose, Mesgul82, Meursault2004, Mgiganteus1, Mic, Micga, Michael Hardy, Michael Zimmermann, Michalws, Mickey gfss2007, Micov, Microtony, Middle Drawer, Mieciu K, Migatu, Mike Rosoft, Miked2144, Mikewebkist, Mild Bill Hiccup, Militaryboy, Milla297, Mimihitam, Minna Sora no Shita, Minpis, Mirek12, Mirv, Misiek889, Missmarple, Mix321, Miyokan, Miztalkalot, Mjanzen, Mjefm, Mnmazur, Modulatum, Molobo, Momo50, Moncrief, Monkeyman, MonoSabio, MoogleEXE, Morwen, Mouse-Touch, Mps, Mr. Chainsaw, MrFish, Msafiri, Mschel, Mushroom, Mvjs, Mx3, Myanw, Mysekurity, Myszodorn, NClement, NHRHS2010, Naddy, Naive cynic, Nakon, NameThatWorks, Naohiro19, Naryathegreat, Nasz, Natalie Erin, Natl1, NawlinWiki, Nayvik, Necronaut, Neferkare, Neurorocker, Neverquick, New4325, NewEnglandYankee, NewbieDoo, Ngchen, Nicegunody, NickW557, Nickfraser, Nicolehere, Nietecza, NigelR, Nightstallion, Nihil novi, Niko226, Nima1024, Niteowlneils, Nivix, Njaelkies Lea, Nksd, Nol888, NorbertArthur, Norm mit, NormanBro, Norum, Notheruser, Notthe600, Nouanoua, NuclearVacuum, NuclearWarfare, Nufy8, Number 57, Numbo3, Nummymuffin, Nv8200p, Nyyankees2fn, Oberst, Obli, Oda Mari, Odie5533, Offa of Angel, Ohnder, Ohnoitsjamie, Oktan, Old Moonraker, Oldgregg13, Oleg Alexandrov, Olga Raskolnikova, Olivier, Oliwier1111, Oly Majoieka, Omenge, Omicronpersei8, Omnibus, Oneiros, Oneness01, Only, Ontheglobe, Opelio, Optimist on the run, Orange5716, Orczar, Oreo Priest, Ottawa4ever, Oxydo, Oxymoron83, P-Chan, PANONIAN, PFHLai, PG-Rated, Pabix, Pacus, Paddyez, Paine Ellsworth, Palosirkka, Pamejudd, Paragon12321, Parhamr, Parrottsvilleelementary, Pat Payne, Patientone, Paul Benjamin Austin, PaulHanson, PaulinSaudi, Paulomazzeirj, PawRon, Pawel bielsko, Pawel z Niepolomic, PawelO, Pawello071, Paweł ze Szczecina, Pawloo, PaxEquilibrium, Pb30, Pedant17, Pel'ican Sh'it, Pensionero, Pernambuko, Peruvianllama, Peter Horn, Peter558, PeterisP, Pethr, Petri Krohn, Phaedrus86, Phenz, Phgao, Philip Trueman, Phillip J, PhnomPencil, Picus viridis, Piepiepiepie2, Pilotguy, Pink!Teen, Piotr p928s, Piotrus, Pirawik, Pitdog, Placebo123, PlushMerga, Pmj, Podruznik, Poeticbent, PolEater, Polacoloco777, Polako, Polaron, PolishMan33, PolishPatriot, Polishtuner9, Politologia, Polski orzel, Polylerus, Porqin, Postdlf, Poszwa, Powertranz, PresN, PrestonH, Programming Hamster, Prusak, Prvc, Przemysł, Przepla, Pseinstein, Pseudo-Richard, Psy guy, Ptmcd, Pwt898, QaviArtilles, Qazwsx13, Qertis, Qagir, Qtoktok, Quadell, Quarma, Quebec99, Quendus, Queson, Quintote, Quizimodo (usurped), Qxz, R Lowry, R9tgokunks, RG2, RGA BALLY, RK, RL0919, Raciel, Radomil, Radzinski, Rajeshm1980, Ran, RapaNui75, Rarelibra, Ravelair, Raven4x4x, Ravenmoon22, Razer rob, RazorICE, Razum2010, Rbarreira, Rd232, Realberserker, RedArmyClub, RedMC, RedWolf, RedWordSmith, ReddShadoe, Redward1, Regulus marzo4103, Regulusmarzo@hotmai.l.com, Rejedef, Remedios44, Retiono Virginian, Retired username, Rettetast, RexNL, Reywas92, Rfl, Rhys, Rich Farmbrough, Richard Harvey, Richardprins, Rick Block, Rickyrab, Rigby27, Rightfully in First Place, Riwnodennyk, Rje, Rjwilmsi, Rklisowski, Rkt2312, Rmhermen, Rmo13, Rmt2m, RoadTrain, Roastytoast, Rob Aleksandrowicz, Rob.derosa, Robert Skyhawk, Robert Weemeyer, Robert346, RobertG, Robertgreer, Robertpolska, Robsuper, Roleplayer, Romann, Romuald Wróblewski, Ronline, Rory096, Roryok, Rowingrocksmysocks, Rreagan007, Rudi Maxer, Ruhrjung, Rumping, Ruok, Ruszewski, Ruud64, Ruy Lopez, Rwerp, RxS, Ryanaxp, Rydia, Ryoho, Ryulong, Ryuu, Rübezahl, SCZenz, SJP, SOPHIAN, SPKirsch, SSJeskimobob, SSZ, SZPANER, Sadi, Sajimem, Salmon L-uh, Salvio giuliano, SamEV, SamWolken, Samah10, Samantha555, Samohyl Jan, Samrolken, Sanbeg, Sango123, Sannse, Sardanaphalus, Satyadasa, Savidan, Sca, Scarletpoet, School is fun, Schopenhauer, Schwartz und Weiss, Science4sail, Scipius, Scohoust, Scott14, Scoutersig, Scwlong, Sean D Martin, Seb az86556, Sebastian770, Sebek2d, Sebjarod, Secretlondon, Seka'sBreathMints, Selfworm, Sengkang, SenorAnderson, Ser Amantio di Nicolao, Seric2, Serte, SexBloodandMetal, Sfahey, Sfu, Shalom Alechem, Shanes, Shearonink, Sheitan, Shield, Shoeofdeath, Shoy, Shuffdog, Shugo255, Sidume, Siemowit, Sietse Snel, Signalhead, Signsolid, Silar, Sillyfolkboy, Simon Peter Hughes, SimonP, Simonxag, Sirius85, Sjwk, Skarl the Drummer, Skatemanman, Skater1123, Skeithkiller, Skier Dude, Skizzik, Skooiman, Skoranka, Skuczera, Skurwiel, Slab stone, Sladen, Slakr, Sławojarek, Slecki19, Sliwers, Smith2006, Smitz, Smokizzy, Smyru, SnappingTurtle, SndrAndrss, Snowfalcon cu, Snowmanradio, Soman, SomeHuman, Someguy1221, Soroush83, Sortior, Soumyasch, South Bay, SpNeo, Space Cadet, Spamtroll, Spicy99, Spitzl, SpookyMulder, Springeragh, SqueakBox, Squids and Chips, Srikeit, Ssolbergj, Stanlavisbad, Starr37, Steel, Stefantastic, Stephenb, Steven Weston, Stickee, StoneProphet, Stressbaby 89, Stroganoff, StuMcDee, Styath, Subtropical-man, Suna, Sunholm, Sunquanliangxiuhao, Supertask, Supertouch, Surowiak, Surtsicna, Suva, Svetovid, SwPawel, SwPaweł2, Swedish fusilier, Swid, SwisterTwister, Sycic, Sylwia Ufnalska, Symane, Syvanen,

Szater, Szczecin, Szopen, TBadger, TFMcQ, TFOWR, TUF-KAT, TaerkastUA, Tandrasz, Targeman, Tassedethe, Tatry, TaulPaul, Tavrian, Taw, Tawker, Taxman, Tbhotch, TeaDrinker, Team Poland, Teddycom, Tekken50, Template namespace initialisation script, Teodor von Zalenzer Halde, Terfili, Terra Novus, Terrencekeenan, Teysz Kamieński, Th' Duck, Th1rt3en, ThaGrind, Thatdog, The Article Creator, The Catholic Knight, The Rambling Man, The sunder king, TheHYPO, TheKMan, Theanthrope, Thehelpfulone, Thelosmets08, Themanwithoutapast, Therequiembellishere, Theroen, Thesoccermage, Thingg, This user has left wikipedia, Thiseye, ThomasK, Thomaspca, Thomehr, Thorsten1, Thryduulf, Thw1309, Tide rolls, TimComm, Timrollpickering, Tiptoety, Titoxd, Tkynerd, Tnick12, Tobby72, Tomaso, Tomasz J Kotarba, Tomirlik, Tommaisey, Tommyjs, Tone, Totti76, TownDown, Tpbradbury, Travelbird, Treisijs, Trekway.com, Trialsanderrors, Tryingtomakethingsright, Tsujigiri, Tuckerresearch, Turmerick, Tuspm, Tvieth01, Tweenk, TycoonTim, TygerDawn, Tymek1988, Typobox, Ulmanor, Ulric1313, Ultratomio, Umedard, Underlying lk, UnicornTapestry, Upi, User86654, Usergreatpower, UtherSRG, Uxejn, Valenciano, Valentinian, Valermos, Vampain, Van helsing, VanCleave, Vaquero100, Vardion, VegaDark, Venariul, VenomousConcept, Ventusa, Viewfinder, Viriditas, Visor, Vitund, Vkem, Vladek Komorek, Von Fiszman, VonShroom, Vortexrealm, Vorthax, Vrinan, Vsmith, WGee, WKS Śląsk Wrocław, WOSlinker, WWEDawn, Waldir, Walton One, Wangi, WegianWarrior, Wehberf, WellsSouth, Weorthe, Werdan7, Werideatdusk33, West Brom 4ever, WhisperToMe, Whomp, Wichid, Wik, Wiki alf, Wikibofh, Wikiborg, Wikidexel2, Wikidudeman, Wikiperuvian, Wilberforcehumphries, Will Beback Auto, William Allen Simpson, William Avery, Wilsonyork, Wimt, Witkacy, Wizzard2k, Wknight94, Woertche.Leo, WojPob, Wojgniew, WolfgangPeters, Wolfrock, Woohookitty, Wtmitchell, Wudkanz, Wulfstan, Wuzzy, Ww2censor, Wyklety, XParadigm777x, XSmartCutie217, Xathaec, Xezbeth, Xlkce, Xtrump, Xx236, YUL89YYZ, Yakudza, Yamaguchi, Yamaka122, Yarovit, Yasis, Yattum, Ybgursey, YellowMonkey, Yonas29, Yonna, Yorkshirian, Youstill let me make accounts lol, Zabadinho, Zanimum, Zaphod Beeblebrox, Zbihniew, Zello, Zenit4ever, ZimZalaBim, Zinjixmaggir, Zippy, Zlerman, Zoney, Zsinj, Zundark, Zzuuzz, Żlepskż, Δ, Александър, 3003 anonymous edits

Bylin *Source*: http://en.wikipedia.org/w/index.php?title=Bylin *Contributors*:

Main_Page *Source*: http://en.wikipedia.org/w/index.php?title=Main_Page *Contributors*: -- April, 0, 130.94.121.xxx, 130.94.122.xxx, 168..., 192.168.32.xxx, 194.82.103.xxx, 200.216.11.xxx, 203.97.100.xxx, 206.230.0.xxx, 210.50.203.xxx, 210.68.39.xxx, 210.8.232.xxx, 217.80.19.xxx, 24.188.157.xxx, 62.252.160.xxx, 63.77.14.xxx, ABCD, Aaron Schulz, Acalamari, Adam Bishop, Ahoerstemeier, Alex756, Alison, Alkivar, AmiDaniel, Andonic, Andre Engels, Andrevan, Anetode, Angela, Animum, Anonymous Dissident, Anthere, AntonioMartin, Arbitrarily0, Arvindn, Arwel Parry, AstroNomer, Astronautics, Aude, AxelBoldt, Baldhur, BanyanTree, Bdesham, Blankfaze, BorgHunter, Brandon, Brian0918, Brion VIBBER, Bryan Derksen, BuickCenturyDriver, CBM, CJ, Cacycle, Camembert, Carey Evans, Celestianpower, CesarB, Cgs, Chadloder, Chris 73, Chris Roy, Chuck SMITH, Chuck Smith, Cimon Avaro, CloudNine, Coren, Cprompt, Crazycomputers, Curps, Cyp, Cyrius, Damas, Damian Yerrick, Dan100, Daniel Quinlan, Danny, Dante Alighieri, Dark Shikari, David Kernow, David Levy, David.Monniaux, DavidLevinson, Davidcannon, Dbachmann, Dbenbenn, Deb, Deckiller, Delirium, Denelson83, Deskana, Dgrant, Docu, Dori, Dragons flight, Dreamyshade, DropDeadGorgias, Duja, Dysprosia, East718, Ed Poor, Ed g2s, Edokter, ElinorD, Eloquence, Enchanter, Enochlau, Evercat, Everyking, Evil saltine, FF2010, Fantasy, FayssalF, Fennec, Feydey, Ffaker, Filiocht, Finlay McWalter, Frazzydee, Fred Bauder, Fredrik, Fuzheado, G-Man, Gadfium, Gaillimh, GayCommunist, Ged UK, General ATO, Graham87, Groogokk, Guanaco, Gurch, Gurubrahma, HJ Mitchell, Hadal, Happy-melon, Haryadoon, Hcatlin, Hemanshu, Henrygb, Hephaestos, HereToHelp, Hersfold, Howcheng, Hyacinth, Ihcoyc, Ike9898, Ilyanep, Infrogmation, Isis, J Di, Jake Nelson, Jamesday, Jan Hidders, Jdforrester, JeLuF, Jeffrey O. Gustafson, Jengod, Jeronimo, JesseW, Jiang, Jimbo Wales, Jimfbleak, John K, John Raftopoulos, JohnOwens, Jpgordon, Jtdirl, Jtkiefer, Juliancolton, Jwrosenzweig, KF, Kaihsu, Kaldari, Karrmann, Kate, Khendon, King of Hearts, Kingturtle, Kosebamse, Koyaanis Qatsi, Krinkle, Ktsquare, Kurykh, Kyorosuke, LFaraone, Larry Sanger, Larry_Sanger, Lee Daniel Crocker, Lexor, LittleDan, Llywrch, Lord Emsworth, Lowellian, Ludraman, Luis Oliveira, Luk, Luna Santin, Lupin, MZMcBride, MacGyverMagic, Magnus Manske, Mailer diablo, Majorly, Malcolm, Malcolm Farmer, Mark, Mark Ryan, MartinHarper, Marudubshinki, Mav, MaxSem, Maxim, Maximus Rex, Mbecker, Menchi, Meno25, Mercury, Merovingian, Merphant, Mets501, Mgiganteus1, Michael Hardy, Michael Snow, Michaelritchie200, Mike Peel, Minesweeper, Mintguy, Mirv, Mkweise, Montrealais, Moreschi, Moriori, Muriel Gottrop, Mysekurity, NSLE, Nakon, Nanobug, Nat, Neutrality, Nickptar, Noldoaran, Notheruser, Nousernamesleft, Nphase, OlEnglish, Oliver Pereira, Olivier, Optim, Optimist on the run, Owen, Ozgod, PFHLai, Pakaran, Patrick, Paul Benjamin Austin, Pcb21, Pepsidrinka, Peter Winnberg, Pharos, Picaroon, Pietdesomere, Pilotguy, Poor Yorick, Prodego, Psy guy, Pudeo, Quercusrobur, Ran, Raul654, Rdsmith4, Redwolf24, Remember the dot, RetiredUser2, RexNL, Rhobite, Riana, Rich Farmbrough, RickK, Robdurbar, Robert Merkel, RobertG, Robin Patterson, RockMFR, Ruud Koot, Ryan Postlethwaite, RyanGerbil10, SQL, ST47, Sade, Sango123, Sannse, Schneelocke, Scipius, Seglea, Seth Ilys, Shreshth91, Shubinator, SimonP, Simondrake, Siroxo, Sj, Sjc, Slowking Man, Slrubenstein, Smith03, Snoyes, Someone else, Stan Shebs, Stephen Gilbert, Stevertigo, Sverdrup, Ta bu shi da yu, Talrias, Tannin, Tariqabjotu, Tarquin, Taw, Tawker, Texture, The Anome, The Blade of the Northern Lights, The Cunctator, The Tom, TheDJ, Thehelpfulone, Tim Starling, TimVickers, Timwi, Titoxd, Toby Bartels, Tom-, Tompagenet, Trialsanderrors, Tristanb, TwoOneTwo, Ugen64, UninvitedCompany, Valley2city, Vancouverguy, Viajero, Vicki Rosenzweig, Violetriga, Viridae, Vzbs34, WhisperToMe, WojPob, Woohookitty, Wouterstomp, YellowMonkey, Zanimum, Zocky, Zoe, Zundark, Zzyzx11, ^demon, Ævar Arnfjörð Bjarmason, 39 anonymous edits

Image Sources, Licenses and Contributors

File:Flag_of_Poland.svg *Source*: http://en.wikipedia.org/w/index.php?title=File:Flag_of_Poland.svg *License*: unknown *Contributors*: Anomie, Mifter

File:Herb Polski.svg *Source*: http://en.wikipedia.org/w/index.php?title=File:Herb_Polski.svg *License*: unknown *Contributors*: Aotearoa, Denelson83, Frombenny, Gustavo Szwedowski de Korwin, Halibutt, Klemen Kocjancic, Madden, Odder, Zscout370, 3 anonymous edits

File:EU-Poland.svg *Source*: http://en.wikipedia.org/w/index.php?title=File:EU-Poland.svg *License*: unknown *Contributors*: User:NuclearVacuum

File:Increase2.svg *Source*: http://en.wikipedia.org/w/index.php?title=File:Increase2.svg *License*: unknown *Contributors*: Sarang

Image:Speakerlink.svg *Source*: http://en.wikipedia.org/w/index.php?title=File:Speakerlink.svg *License*: unknown *Contributors*: Woodstone. Original uploader was Woodstone at en.wikipedia

File:Zaprowadzenie chrzescijanstwa 965 Matejko.JPG *Source*: http://en.wikipedia.org/w/index.php?title=File:Zaprowadzenie_chrzescijanstwa_965_Matejko.JPG *License*: unknown *Contributors*: BurgererSF, Goldfritha, Kjetil r, Ludmiła Pilecka, Piotrus, Shakko, Wst, 2 anonymous edits

File:Jan Matejko-Astronomer Copernicus-Conversation with God.jpg *Source*: http://en.wikipedia.org/w/index.php?title=File:Jan_Matejko-Astronomer_Copernicus-Conversation_with_God.jpg *License*: unknown *Contributors*: Alaniaris, BurgererSF, Dirk Hünniger, EugeneZelenko, Gnesener1900, Goldfritha, Kürschner, Ludmiła Pilecka, Matthead, Olivier2, Piotrus, Pko, Plindenbaum, Slomox, Wames, Wst, 3 anonymous edits

File:Kraków - Wawel from Kopiec Krakusa.jpg *Source*: http://en.wikipedia.org/w/index.php?title=File:Kraków_-_Wawel_from_Kopiec_Krakusa.jpg *License*: unknown *Contributors*: User:RaNo

File:Rzeczpospolita 1600.png *Source*: http://en.wikipedia.org/w/index.php?title=File:Rzeczpospolita_1600.png *License*: unknown *Contributors*: Bandar Lego, Dcoetzee, Denniss, Electionworld, It Is Me Here

File:SiemiginowskiJerzy.1686.JanIIISobieskiPodWiedniem.jpg *Source*: http://en.wikipedia.org/w/index.php?title=File:SiemiginowskiJerzy.1686.JanIIISobieskiPodWiedniem.jpg *License*: unknown *Contributors*: AnonMoos, BurgererSF, Gryffindor, Infrogmation, Piotrus, Rotational, Shalom Alechem, Umherirrender, Wames, 1 anonymous edits

File:Stanisław II August Poniatowski in coronation clothes.PNG *Source*: http://en.wikipedia.org/w/index.php?title=File:Stanisław_II_August_Poniatowski_in_coronation_clothes.PNG *License*: unknown *Contributors*: BurgererSF, Ecummenic, Kürschner, Mathiasrex

File:Smuglewicz Kosciuszko 2.jpg *Source*: http://en.wikipedia.org/w/index.php?title=File:Smuglewicz_Kosciuszko_2.jpg *License*: unknown *Contributors*: Alma Pater, DieBuche, DocentX, Goldfritha, Man vyi, Mathiasrex, Sfu, Wames, 8 anonymous edits

File:Starcie belwederczykow z kirasjerami rosyjskimi na moscie w Lazienkach.jpg *Source*: http://en.wikipedia.org/w/index.php?title=File:Starcie_belwederczykow_z_kirasjerami_rosyjskimi_na_moscie_w_Lazienkach.jpg *License*: unknown *Contributors*: Allen3, Bedford, Emax, Magog the Ogre, Piotrus, Raven4x4x

File:Jozef Pilsudski.jpg *Source*: http://en.wikipedia.org/w/index.php?title=File:Jozef_Pilsudski.jpg *License*: unknown *Contributors*: Alma Pater, Ejdzej, EugeneZelenko, Kocio, Mathiasrex, Serdelll, 1 anonymous edits

File:Rzeczpospolita 1939.svg *Source*: http://en.wikipedia.org/w/index.php?title=File:Rzeczpospolita_1939.svg *License*: unknown *Contributors*: User:Halibutt

File:10Dyw.JPG *Source*: http://en.wikipedia.org/w/index.php?title=File:10Dyw.JPG *License*: unknown *Contributors*: Andros64, GeorgHH, Jarekt, Mathiasrex, PMG, Pibwl, Polish29, Robek, Sandmann4u, Thib Phil, 3 anonymous edits

File:Curzon line en.svg *Source*: http://en.wikipedia.org/w/index.php?title=File:Curzon_line_en.svg *License*: unknown *Contributors*: User:radek.s

File:Polish Soldier's Grave Warsaw 1945.jpg *Source*: http://en.wikipedia.org/w/index.php?title=File:Polish_Soldier's_Grave_Warsaw_1945.jpg *License*: unknown *Contributors*: w:pl:Tadeusz BukowskiTadeusz Bukowski

File:Bundesarchiv Bild 183-F0417-0001-011, Berlin, VII. SED-Parteitag, Eröffnung.jpg *Source*: http://en.wikipedia.org/w/index.php?title=File:Bundesarchiv_Bild_183-F0417-0001-011,_Berlin,_VII._SED-Parteitag,_Eröffnung.jpg *License*: unknown *Contributors*: Kohls, Ulrich

File:Solidarity poster 1989.jpg *Source*: http://en.wikipedia.org/w/index.php?title=File:Solidarity_poster_1989.jpg *License*: unknown *Contributors*: Andy Dingley, BurgererSF

File:Flaga RP z UE.jpg *Source*: http://en.wikipedia.org/w/index.php?title=File:Flaga_RP_z_UE.jpg *License*: unknown *Contributors*: Michal Osmenda

File:Polen topo.jpg *Source*: http://en.wikipedia.org/w/index.php?title=File:Polen_topo.jpg *License*: unknown *Contributors*: Antemister, Captain Blood, Hystrix, 2 anonymous edits

File:Śląskie Kamienie - Dívči Kameny.JPG *Source*: http://en.wikipedia.org/w/index.php?title=File:Śląskie_Kamienie_-_Dívči_Kameny.JPG *License*: unknown *Contributors*: User:Pudelek

File:Giewont and Wielka Turnia.jpg *Source*: http://en.wikipedia.org/w/index.php?title=File:Giewont_and_Wielka_Turnia.jpg *License*: unknown *Contributors*: Jerzy Opioła

File:2 SPN 01.jpg *Source*: http://en.wikipedia.org/w/index.php?title=File:2_SPN_01.jpg *License*: unknown *Contributors*: BLueFiSH.as, Merlin, Panek, Piotrus, Reytan, Shalom Alechem, Uroboros, 2 anonymous edits

File:Modlin spichlerz.jpg *Source*: http://en.wikipedia.org/w/index.php?title=File:Modlin_spichlerz.jpg *License*: unknown *Contributors*: User:Wojsyl

File:Kurtkowiec i czerwone.jpg *Source*: http://en.wikipedia.org/w/index.php?title=File:Kurtkowiec_i_czerwone.jpg *License*: unknown *Contributors*: Adi, Adziura, Cathy Richards, Daniel Baránek, Dome, Michau Sm, Panek, Selso, Shalom Alechem, ToSter, Ulamm, 5 anonymous edits

File:Solina Poland.jpg *Source*: http://en.wikipedia.org/w/index.php?title=File:Solina_Poland.jpg *License*: unknown *Contributors*: Man vyi, Segu, Shalom Alechem, Silar, 2 anonymous edits

File:Zatoka Pucka - Bay of Puck (8).jpg *Source*: http://en.wikipedia.org/w/index.php?title=File:Zatoka_Pucka_-_Bay_of_Puck_(8).jpg *License*: unknown *Contributors*: Krzysztof, Silar, 1 anonymous edits

File:Gora Cisowa 03.jpg *Source*: http://en.wikipedia.org/w/index.php?title=File:Gora_Cisowa_03.jpg *License*: unknown *Contributors*: Jarekt, Michau Sm, Shalom Alechem, 7 anonymous edits

File:WhiteStorkFamily.jpg *Source*: http://en.wikipedia.org/w/index.php?title=File:WhiteStorkFamily.jpg *License*: unknown *Contributors*: User:Manfred Heyde

File:Wisent.jpg *Source*: http://en.wikipedia.org/w/index.php?title=File:Wisent.jpg *License*: unknown *Contributors*: Henryk Kotowski,

File:Plaża w Pucku - kitesurfers - beach in Puck (4).jpg *Source*: http://en.wikipedia.org/w/index.php?title=File:Plaża_w_Pucku_-_kitesurfers_-_beach_in_Puck_(4).jpg *License*: unknown *Contributors*: Fagairolles 34, Krzysztof, Marek Banach, 3 anonymous edits

File:Bronisław Komorowski rotcropped.jpg *Source*: http://en.wikipedia.org/w/index.php?title=File:Bronisław_Komorowski_rotcropped.jpg *License*: unknown *Contributors*: User:Saibo

File:PL Sejm hall.jpg *Source*: http://en.wikipedia.org/w/index.php?title=File:PL_Sejm_hall.jpg *License*: unknown *Contributors*: User:Network.nt

File:Polishsupremecourt.JPEG *Source*: http://en.wikipedia.org/w/index.php?title=File:Polishsupremecourt.JPEG *License*: unknown *Contributors*: Original uploader was Cjs2111 at en.wikipedia

File:Manuscript of the Constitution of the 3rd May 1791.PNG *Source*: http://en.wikipedia.org/w/index.php?title=File:Manuscript_of_the_Constitution_of_the_3rd_May_1791.PNG *License*: unknown *Contributors*: Anonymous

File:Radoslaw Sikorski meets Secretary Hillary Clinton.jpg *Source*: http://en.wikipedia.org/w/index.php?title=File:Radoslaw_Sikorski_meets_Secretary_Hillary_Clinton.jpg *License*: unknown *Contributors*: Michael Gross, State Department photographer

File:Walesa and Tusk EPP.jpg *Source*: http://en.wikipedia.org/w/index.php?title=File:Walesa_and_Tusk_EPP.jpg *License*: unknown *Contributors*: europeanpeoplesparty

Image:POL location map.svg *Source*: http://en.wikipedia.org/w/index.php?title=File:POL_location_map.svg *License*: unknown *Contributors*: Hiuppo

Image:POL województwo zachodniopomorskie COA.svg *Source*: http://en.wikipedia.org/w/index.php?title=File:POL_województwo_zachodniopomorskie_COA.svg *License*: unknown *Contributors*: User:Bastianow

Image:POL województwo pomorskie COA.svg *Source*: http://en.wikipedia.org/w/index.php?title=File:POL_województwo_pomorskie_COA.svg *License*: unknown *Contributors*: User:Bastianow

Image:Warminsko-mazurskie herb.svg *Source*: http://en.wikipedia.org/w/index.php?title=File:Warminsko-mazurskie_herb.svg *License*: unknown *Contributors*: Derbeth, Frombenny, Gustavo Szwedowski de Korwin, JDavid, MAC13, Serdelll, 1 anonymous edits

Image:POL województwo podlaskie COA.svg *Source*: http://en.wikipedia.org/w/index.php?title=File:POL_województwo_podlaskie_COA.svg *License*: unknown *Contributors*: Bastianow, Derbeth, Frombenny, Gustavo Szwedowski de Korwin, JDavid, WarX, XtraVert, 2 anonymous edits

Image:POL województwo lubuskie COA.svg *Source*: http://en.wikipedia.org/w/index.php?title=File:POL_województwo_lubuskie_COA.svg *License*: unknown *Contributors*: User:Bastianow

Image:POL województwo wielkopolskie COA.svg *Source*: http://en.wikipedia.org/w/index.php?title=File:POL_województwo_wielkopolskie_COA.svg *License*: unknown *Contributors*: User:Bastianow

File:Ratusz2007.jpg *Source*: http://en.wikipedia.org/w/index.php?title=File:Ratusz2007.jpg *License*: unknown *Contributors*: adamchemik

File:The tall ships races.jpg *Source*: http://en.wikipedia.org/w/index.php?title=File:The_tall_ships_races.jpg *License*: unknown *Contributors*: Nanc1 Original uploader was Nanc1 at pl.wikipedia

File:Kotlet schabowy.jpg *Source*: http://en.wikipedia.org/w/index.php?title=File:Kotlet_schabowy.jpg *License*: unknown *Contributors*: phototram

file:Poland location map.svg *Source*: http://en.wikipedia.org/w/index.php?title=File:Poland_location_map.svg *License*: unknown *Contributors*: User:NordNordWest

File:Red pog.svg *Source*: http://en.wikipedia.org/w/index.php?title=File:Red_pog.svg *License*: unknown *Contributors*: Anomie

File:Flag of Poland.svg *Source*: http://en.wikipedia.org/w/index.php?title=File:Flag_of_Poland.svg *License*: unknown *Contributors*: Anomie, Mifter

File:Mojo Mathers.jpg *Source*: http://en.wikipedia.org/w/index.php?title=File:Mojo_Mathers.jpg *License*: unknown *Contributors*: Photographer unidentified, Green Party NZ is copyright holder.

File:ThreeGorgesDam-China2009.jpg *Source*: http://en.wikipedia.org/w/index.php?title=File:ThreeGorgesDam-China2009.jpg *License*: unknown *Contributors*: User:Rehman

File:Three Sisters Sunset.jpg *Source*: http://en.wikipedia.org/w/index.php?title=File:Three_Sisters_Sunset.jpg *License*: unknown *Contributors*: User:JJ Harrison

Printed by Books on Demand GmbH, Norderstedt / Germany